AF778184

Easy Journey to Other Planets

BOOKS by
His Divine Grace A. C. Bhaktivedanta Swami Prabhupāda

Bhagavad-gītā As It Is
Śrīmad-Bhāgavatam (completed by disciples)
Śrī Caitanya-caritāmṛta
Kṛṣṇa, the Supreme Personality of Godhead
Teachings of Lord Caitanya
The Nectar of Devotion
The Nectar of Instruction
Śrī Īśopaniṣad
Light of the Bhāgavata
Easy Journey to Other Planets
Teachings of Lord Kapila, the Son of Devahūti
Teachings of Queen Kuntī
Message of Godhead
The Science of Self-Realization
The Perfection of *Yoga*
Beyond Birth and Death
On the Way to Kṛṣṇa
Rāja-vidyā: The King of Knowledge
Elevation to Kṛṣṇa Consciousness
Kṛṣṇa Consciousness: The Matchless Gift
Kṛṣṇa Consciousness: The Topmost *Yoga* System
Perfect Questions, Perfect Answers
Life Comes from Life
The Nārada-bhakti-sūtra (completed by disciples)
The Mukunda-mālā-stotra (completed by disciples)
Geetār-gān (Bengali)
Vairāgya-vidyā (Bengali)
Buddhi-*yoga* (Bengali)
Bhakti-ratna-boli (Bengali)
Back to Godhead magazine (founder)

BOOKS compiled from the teachings of
His Divine Grace A. C. Bhaktivedanta Swami Prabhupāda
after his lifetime

Search for Liberation
A Second Chance
The Journey of Self-Discovery
Civilization and Transcendence
The Laws of Nature
Renunciation Through Wisdom
The Quest for Enlightenment
Dharma, the Way of Transcendence
Beyond Illusion and Doubt
Bhakti-yoga: The Art of Eternal Love
Spiritual Yoga

Easy Journey to Other Planets

HIS DIVINE GRACE
A. C. Bhaktivedanta Swami Prabhupāda
Founder-*Ācārya* of the International Society for Krishna Consciousness

THE BHAKTIVEDANTA BOOK TRUST
Los Angeles • Stockholm • Mumbai • Sydney

Readers interested in the subject matter of this book are invited by the International Society for Krishna Consciousness to correspond with its secretary.

International Society for Krishna Consciousness
P.O. Box 341445
Los Angeles, California 90034, USA
Telephone: 1-800-927-4152 (Inside USA);
1-310-837-5283 (Outside USA)
Fax: 1-310-837-1056
e-mail: bbt.usa@krishna.com
web: www.krishna.com

International Society for Krishna Consciousness
P.O. Box 380
2765 Riverstone NSW, Australia
Phone: +61-2-9627-6306
Fax: +61-2-9627-6052
E-mail: bbt.wp@krishna.com

Previous Printings: 2,300,000
This Printing, February 2022: 100,000

© 1972 The Bhaktivedanta Book Trust International, Inc.

All Rights Reserved

Printed in China

ISBN 0-912776-10-2

Dedicated to
the scientists of the world,
with blessings of
His Divine Grace
Śrī Śrīmad Bhaktisiddhānta Sarasvatī
Gosvāmī Mahārāja,
my spiritual master

Contents

Preface

A living being, especially civilized man, has a natural desire to live forever in happiness. This is quite natural because, in his original state, the living being is both eternal and joyful. However, in the present conditioned state of life, he is engaged in a struggle against recurring birth and death. Therefore he has attained neither happiness nor immortality.

The latest desire man has developed is the desire to travel to other planets. This is also quite natural, because he has the constitutional right to go to any part of the material or spiritual skies. Such travel is very tempting and exciting because these skies are full of unlimited globes of varying qualities, and they are occupied by all types of living entities. The desire to travel there can be fulfilled by the process of *yoga*, which serves as a means by which one can transfer himself to whatever planet he likes—possibly to planets where life is not only eternal and blissful, but where there are multiple varieties of enjoyable energies. Anyone who can attain the freedom of the spiritual planets need never return to this miserable land of birth, old age, disease and death.

One can attain this stage of perfection very easily by his individual effort. He can simply follow, in his own home, the prescribed method of *bhakti-yoga*. This method, under proper guidance, is simple and enjoyable. An attempt is made herein to give information to the people in general, and to philosophers and religionists in particular, as to how one can transfer oneself to other planets by this process of *bhakti-yoga*—the highest of all yogic processes.

1

Antimaterial Worlds

Materialistic science may one day finally discover the eternal antimaterial world, which has for so long been unknown to the wranglers of gross materialism. Regarding the scientists' present conception of antimatter, the *Times of India* (Oct. 27, 1959) published the following news release:

> *Stockholm, Oct. 26, 1959* — Two American atomic scientists were awarded the 1959 Nobel Physics Prize today for the discovery of the antiproton, proving that matter exists in two forms—as particles and antiparticles. They are Italian—born Dr. Emillo Segre, 69, and Dr. Owen Chamberlain, born in San Francisco. . . . According to one of the fundamental assumptions of the new theory, there may exist another world, or an anti-world, built up of antimatter. This antimaterial world would consist of atomic and subatomic particles spinning in reverse orbits to those of the

> world we know. If these two worlds should ever clash, they would both be annihilated in one blinding flash.

In this statement, the following propositions are put forward:

1. There is an antimaterial atom or particle which is made up of the anti-qualities of material atoms.
2. There is another world besides this material world, of which we have only limited experience.
3. The antimaterial and material worlds may clash at a certain period and may annihilate one another.

Out of these three items, we, the students of theistic science, can fully agree with items 1 and 2, but we can agree with item 3 only within the limited scientific definition of antimatter. The difficulty lies in the fact that the scientists' conception of antimatter extends only to another variety of material energy, whereas the real antimatter must be entirely antimaterial. Matter as it is constituted is subjected to annihilation, but antimatter—if it is to be free from all material symptoms—must also be free from annihilation, by its very nature. If matter is destructible or separable, antimatter must be indestructible and inseparable. We shall try to discuss these propositions from the angle of authentic scriptural vision.

The most widely recognized scriptures in the world are the *Vedas*. The *Vedas* have been divided into four parts: *Sāma*, *Yajur*, *Ṛg* and *Atharva*. The subject matter of the *Vedas* is very difficult for a man of ordinary understanding. For elucidation, the four *Vedas* are explained in the historical epic called the *Mahābhārata* and in eighteen *Purāṇas*. The *Rāmāyaṇa* is also a historical epic which contains all the necessary information from the *Vedas*. So the four *Vedas*, the original *Rāmāyaṇa* by Vālmīki, the *Mahābhārata* and the *Purāṇas* are classified as Vedic literatures. The *Upaniṣads* are parts of the four *Vedas*, and the *Vedānta-sūtras* represent the cream of the *Vedas.* To summarize all these Vedic literatures, the *Bhagavad-gītā* is accepted as the essence of all *Upaniṣads* and the preliminary explanation of the *Vedānta-sūtras.* One may then conclude that from the *Bhagavad-gītā* alone one can have the essence of the *Vedas,* for it is spoken by Lord Śrī Kṛṣṇa, the Supreme Personality of Godhead, who descends to this material world from the antimaterial world in order to give complete information of the superior form of energy.

The superior form of energy of the Personality of Godhead is described in the *Bhagavad-gītā* as *parā prakṛti.* The scientists have recently discovered that there are two forms of perishable matter, but the *Bhagavad-gītā* describes most perfectly the concept of matter and antimatter in terms of two forms of energy. Matter is an energy which creates the material world, and the same energy, in its superior form, also creates the antimaterial (transcendental) world. The living entities belong to the category of superior energy. The inferior energy, or material energy, is called *aparā prakṛti.* In the *Bhagavad-gītā* the creative energy is thus presented in two forms, namely *aparā* and *parā prakṛti.*

Matter itself has no creative power. When it is manipulated by the living energy, material things are produced. Matter in its crude form is therefore the latent energy of the Supreme Being. Whenever we think of energy, it is natural that we think of the source of energy. For example, when we think of electrical energy, we simultaneously think of the powerhouse where it is generated. Energy is not self-sufficient. It is under the control of a superior living being. For example, fire is the source of two other energies, namely light and heat. Light and heat have no independent existence outside of fire. Similarly, the inferior and superior energies are derived from a source, which one may call by any name. That source of energy must be a *living being* with full sense of everything. That supreme living being is the Personality of Godhead, Śrī Kṛṣṇa, or the all-attractive living being.

In the *Vedas* the supreme living being, or the Absolute Truth, is called Bhagavān—the opulent one, the living being who is the fountainhead of all energies. The discovery of the two forms of limited energies by the modern scientists is just the beginning of the progress of science. Now they must go further to discover the source of the two particles or atoms which they term material and antimaterial.

How can the antimaterial particle be explained? We have experience with material particles or atoms, but we have no experience with antimaterial atoms. However, the *Bhagavad-gītā* gives the following vivid description of the antimaterial particle:

> "This antimaterial particle is within the material body. Because of the presence of this antimaterial particle, the material body is progressively changing from childhood to boyhood, from boyhood to youth to old age, after which the antimaterial particle leaves the old, unworkable body and takes up another material body."

This description of a living body confirms the scientific discovery that energy exists in two forms. When one of them, the antimaterial particle, is separated from the material body, the latter becomes useless for all purposes. As such, the antimaterial particle is undoubtedly superior to the material energy.

> "No one, therefore, should lament for the loss of material energy. All varieties of sense perception in the categories of heat and cold, happiness and distress, are but interactions of material energy which come and

> go like seasonal changes. The temporary appearance and disappearance of such material interactions confirms that the material body is formed of a material energy inferior to the living force, or *jīva* energy."
>
> "Any intelligent man who is not disturbed by happiness and distress, understanding that they are different material phases resulting from the interactions of the inferior energy, is competent to regain the antimaterial world, where life is eternal, full of permanent knowledge and bliss."

The antimaterial world is mentioned here, and in addition information is given that in the antimaterial world there is no "seasonal" fluctuation. Everything there is permanent, blissful, and full of knowledge. But when we speak of it as a "world," we must remember that it has forms and paraphernalia of various categories beyond our material experiences.

> "The material body is destructible, and as such it is changeable and temporary. So is the material world. But the antimaterial living force is nondestructible, and therefore it is permanent. Expert scientists have thus distinguished the different qualities of the material and antimaterial particles as temporary and permanent respectively."

The discoverers of the two forms of matter have yet to find out the qualities of antimatter. But a vivid description is already given in the *Bhagavad-gītā* as follows. The scientist can make further research on the basis of this valuable information.

> "The antimaterial particle is finer than the finest of material particles. This living force is so powerful that

> it spreads its influence all over the material body. The antimaterial particle has immense potency in comparison to the material particle, and consequently it cannot be destroyed."

This is but the beginning of the description of the antimaterial particle in the *Bhagavad-gītā*. It is further explained as follows:

> "The finest form of the antimaterial particle is encaged within the gross and subtle material bodies. Although the material bodies (both gross and subtle) are subject to destruction, the finer, antimaterial particle is eternal. One's interest, therefore, should be in this eternal principle."

The perfection of science will occur when it is possible for the material scientists to know the qualities of the antimaterial particle and liberate it from the association of nonpermanent, material particles. Such liberation would mark the culmination of scientific progress.

There is partial truth in the scientists' suggestion that there may exist also another world consisting of antimaterial atoms and that a clash between the material and antimaterial worlds will result in the annihilation of both. There is a clash which is continually going on: the annihilation of the material particles is taking place at every moment, and the nonmaterial particle is striving for liberation. This is explained in the *Bhagavad-gītā* as follows:

> "The nonmaterial particle, which is the living entity, influences the material particle to work. This living entity is always indestructible. As long as the nonmaterial particle is within the lump of material

> energy—known by the names of gross and subtle bodies—then the entity is manifest as a living unit. In the continuous clashing between the two particles, the nonmaterial particle is never annihilated. No one can destroy the antimaterial particle at any time—past, present or future."

Therefore, we think that the theory maintaining that the material and antimaterial worlds may clash, resulting in the annihilation of both worlds, is correct only within the context of the scientists' limited definition of antimatter. The *Bhagavad-gītā* explains the nature of the antimaterial particle, which can never be annihilated:

> "The fine and immeasurable antimaterial particle is always indestructible, permanent and eternal. After a certain period, however, its encagement by material particles is annihilated. This same principle also operates in the case of the material and antimaterial worlds. No one should fear the annihilation of the antimaterial particle, for it survives the annihilation of material worlds."

Everything that is created is annihilated at a certain stage. Both the material body and the material world are created, and they are therefore subject to annihilation. The antimaterial particle, however, is never created, and consequently it is never annihilated. This also is corroborated in the *Bhagavad-gītā:*

> "The antimaterial particle, which is the vital force, is never born or created. It exists eternally. It has neither birth dates nor death dates. It is neither repeatedly created nor repeatedly destroyed. It is eternally

> existing, and therefore it is the oldest of the old, and yet it is always fresh and new. Although the material particle is annihilated, the antimaterial particle is never affected."

The principle is also applicable to the antimaterial universe as well as to the antimaterial particle. When the material universe is annihilated, the antimaterial universe exists in all circumstances. This will be explained in more detail later.

The scientist may also learn the following from the *Bhagavad-gītā:*

> "The learned man who knows perfectly well that the antimaterial particle is indestructible knows that it cannot be annihilated by any means."

The atomic scientist may consider annihilating the material world by nuclear weapons, but his weapons cannot destroy the antimaterial world. The antimaterial particle is more clearly explained in the following lines:

> "It is neither cut into pieces by any material weapon, nor is it burnt by fire. Nor is it moistened by water, nor withered, nor dried up, nor evaporated in the air. It is indivisible, nonflammable and insoluble. Because it is eternal, it can enter into and leave any sort of body. Being steady by constitution, its qualities are always fixed. It is inexplicable, because it is contrary to all material qualities. It is unthinkable by the ordinary brain. It is unchangeable. No one, therefore, should ever lament for what is an eternal, antimaterial principle."

Thus, in the *Bhagavad-gītā* and in all other Vedic literatures the superior energy (antimaterial principle) is accepted as the vital force, or the living spirit. This is also called the *jīva*. This living principle cannot be generated by any combination of material elements. There are eight material principles which are described as inferior energies, and they are: (1) earth, (2) water, (3) fire, (4) air, (5) ether, (6) mind, (7) intelligence and (8) ego. Apart from these is the living force, or the antimaterial principle, which is described as the superior energy. These are called "energies" because they are wielded and controlled by the supreme living being, the Personality of Godhead (Kṛṣṇa).

For a long time the materialist was limited within the boundary of the eight material principles mentioned above. Now it is encouraging to see that he has a little preliminary information of the antimaterial principle and the anti-material universe. We hope that with the progress of time the materialist will be able to estimate the value of the antimaterial world, in which there is no trace of material principles. Of course the very word "antimaterial" indicates that the principle is in opposition to all material qualities.

There are, of course, the mental speculators who comment upon the antimaterial principle. These fall into two main groups, and they arrive at two different erroneous conclusions. One group (the gross materialists) either denies the antimaterial principle or admits only the disintegration of material combination at a certain stage (death). The other group accepts the antimaterial principle as being in direct opposition to the material principle with its twenty-four categories. This group is known as the Sāṅkhyaites, and they investigate the material principles and analyze them minutely. At the end of their investigation, the Sāṅkhyaites finally accept only a transcendental (antimaterial) nonactive principle. However, difficulties arise for all these mental speculators because they speculate with the help of inferior energy. They do not accept information from the superior. In order to realize the real position of the antimaterial principle, one must rise to the transcendental plane of superior energy. *Bhakti-yoga* is the very activity of superior energy.

From the platform of the material world, one cannot estimate the real position of the antimaterial world. But the Supreme Lord, who is the controller of both material and antimaterial energies, descends out of His causeless mercy and gives us complete information of the antimaterial world. In this way we can know what the antimaterial world is. *The Supreme Lord and the living entities are both antimaterial in quality,* we are informed. Thus, we can have an idea of the Supreme Lord by an elaborate study of the living entities. Every living entity is an individual person. Therefore, the supreme living being must also be the supreme person. In the Vedic literatures the supreme person is properly claimed to be Kṛṣṇa. The name "Kṛṣṇa," indicating the Supreme Lord, is the only truly intelligible name of the highest order. He is the controller of both material and antimaterial energies, and the very word "Kṛṣṇa" signifies that He is the supreme controller. In the *Bhagavad-gītā* the Lord confirms this as follows:

> "There are two worlds—the material and antimaterial. The material world is composed of inferior qualitative energy divided into eight material principles. The antimaterial world is made of superior qualitative energy. Because both the material and antimaterial energies are emanations of the Supreme Transcendence, the Personality of Godhead, it is proper to conclude that I [Lord Kṛṣṇa] am the ultimate cause of all creations and annihilations."

Because the Lord's two energies (inferior and superior)

manifest the material and antimaterial worlds, He is called the Supreme Absolute Truth. Lord Kṛṣṇa explains this in the *Bhagavad-gītā* thus:

> "I am, Arjuna, the highest principle of transcendence, and there is nothing greater than Me. Everything that be rests on My energies exactly like pearls on a thread."

Long, long before the discovery of the principles of antimatter and the antimaterial worlds, the subject was delineated in the pages of the *Bhagavad-gītā.* The *Gītā* itself indicates that its philosophy had previously been taught to the presiding deity of the sun, which implies that the principles of the *Bhagavad-gītā* were expounded by the Personality of Godhead long before the Battle of Kurukṣetra—at least some 120,000,000 years before. Now modern science has just discovered a fraction of the truths that are available in the *Bhagavad-gītā.*

The assumption of an antimaterial universe is also found in the *Bhagavad-gītā.* And from all data available it is to be assumed without the slightest doubt that the antimaterial world is situated in the antimaterial sky, a sky which is mentioned in the *Bhagavad-gītā* as *sanātana-dhāma,* or the eternal nature.

Exactly as material atoms create the material world, the antimaterial atoms create the antimaterial world with all its paraphernalia. The antimaterial world is inhabited by antimaterial living beings. In the antimaterial world there is no inert matter. *Everything there is a living principle,* and the Supreme Personality in that region is God Himself. The denizens of the antimaterial world possess eternal life, eternal knowledge and eternal bliss. In other words, they have all the qualifications of God.

In the material world the topmost planet is called Satyaloka, or Brahmaloka. Beings of the greatest talents live on this planet. The presiding deity of Brahmaloka is Brahmā, the first created being of this material world. Brahmā is a living being like so many of us, but he is the most talented personality in the material world. He is not so talented that he is in the category of God, but he is in the category of those living entities directly dominated by God. God and the living entities both belong to the antimaterial world. The scientist, therefore, would be rendering service to everyone by researching the constitution of the antimaterial world—how it is administered, how things are shaped there, who are the presiding personalities, and so on. Of the Vedic literatures, *Śrīmad-Bhāgavatam* deals elaborately with these matters. The *Bhagavad-gītā* is the preliminary study of the *Śrīmad-Bhāgavatam*. These two important books of knowledge should be thoroughly studied by all men in the scientific world. These books would give many clues to scientific progress and would indicate many new discoveries.

The transcendentalists and the materialists are two distinct classes of men. The transcendentalist gathers knowledge from authoritative scriptures like the *Vedas*. Vedic literature is received from authoritative sources which are in the line of transcendental disciplic succession. This disciplic succession (*paramparā*) is also mentioned in the *Bhagavad-gītā*. Kṛṣṇa says in the *Bhagavad-gītā* that hundreds of thousands of years ago the *Gītā* was spoken to the presiding deity of the sun, who delivered the knowledge to his son Manu, from whom the present generation of man has descended. Manu, in his turn, delivered this transcendental knowledge to his son King Ikṣvāku, who is the forefather of the dynasty in which the Personality of Godhead Śrī Rāma appeared. This long chain of disciplic succession was broken during the advent period of Lord Kṛṣṇa (five thousand years ago), and for this reason Kṛṣṇa restated the *Bhagavad-gītā* to Arjuna, thereby making him the first disciple of this knowledge in this age. The transcendentalist of this age, therefore, is in the disciplic line that starts with Arjuna. Without troubling himself with materialistic research work, the transcendentalist acquires the truths concerning matter and antimatter in the most perfect way (through this disciplic succession) and thereby saves himself much botheration.

The gross materialists, however, do not believe in the antimaterial worlds of the Personality of Godhead. They are therefore unfortunate creatures, although sometimes very talented, educated and advanced otherwise. They are bewildered by the influence of the material manifestation and are devoid of knowledge of things antimaterial. It is a good sign, therefore, that the materialistic scientists are gradually progressing toward the region of the antimaterial world. It may even be possible for them to make sufficient progress to be able to know the details of this antimaterial world, where the Personality of Godhead resides as the predominating figure and where the living entities live with Him and serve Him. The living entities who serve the Godhead are equal in quality to Him, but at the same time they are predominated as servitors. In the antimaterial world there is no difference between the predominated and the predominator—the relationship is in perfection and without tinge of materialism.

The nature of the material world is destructive. According to the *Bhagavad-gītā,* there is some partial truth to the assumption of the physical scientist that there is annihilation of the material and antimaterial worlds when they chance to clash. The material world is a creation of changing modes of nature. These modes (*guṇas*) are known as *sattva* (goodness), *rajas* (passion) and *tamas* (ignorance). The material world is created by the mode of *rajas,* maintained by the mode of *sattva,* and annihilated by the mode of *tamas.* These modes are omnipresent in the material world, and as such, at every hour, every minute, every second, the process of creation, maintenance and annihilation is taking place all over the material universe. The highest planet of the material universe, Brahmaloka, is also subjected to these modes of nature, although the duration of life on that planet, due to the predominance of the mode of *sattva,* is said to be 4,300,000 x 1,000 x 2 x 30 x 12 x 100 solar years. Despite this long duration, however, Brahmaloka is subject to destruction. Although life on Brahmaloka is fantastically long compared to life on Earth, it is only a flash in comparison to the eternal life of the nonmaterial worlds. Consequently, the speaker of the *Bhagavad-gītā,* Lord Śrī Kṛṣṇa, asserts the importance of the antimaterial universe, which is His abode.

> "All the planets within the material universe are destroyed at the end of 4,300,000 x 1,000 x 2 x 30 x 12 x 100 solar years. And all the living beings inhabiting these material planets are destroyed materially along with the destruction of the material worlds. The living entity, however, is constitutionally an antimaterial particle. But unless he elevates himself to the region of the antimaterial worlds by cultivation of antimaterial activities, he is destroyed materially at the annihilation of the material worlds and is subject to take rebirth in a material shape with the rebirth of a new material universe. In other words, he is subject to the pains of repeated birth and death. *Only those living entities who take to the loving service of the Personality of Godhead during the manifested stage of material life are undoubtedly transferred to the antimaterial worlds after quitting the material body.* Immortality is obtained only by those who return to

Godhead by practice of antimaterial activities."

What are these antimaterial activities? They are medicines. For example, when a man falls ill, he goes to a physician who prescribes medicines which eventually cure the suffering patient. Similarly, the materialist is ailing, and he should consult an expert transcendentalist-physician. What is his ailment? He is suffering the tribulations of repeated births, deaths, diseases and old age. Once he agrees to put himself under the "back to Godhead" treatment, he is able to transfer himself to the antimaterial world, where there is eternal life instead of birth and death.

Annihilation of the material world takes place in two ways. Partial annihilation occurs at the end of every 4,300,000 x 1,000 solar years, or at the end of each day of Brahmaloka, which is the topmost planet in the material world. During that time of partial annihilation, the topmost planets such as Brahmaloka are not annihilated, but at the end of each duration of 4,300,000 x 1,000 x 2 x 30 x 12 x 100 solar years, the entire cosmic manifestation is merged into the antimaterial body, from whence the material principles had emanated and become manifest. The antimaterial world, which is far removed from the material sky, is never annihilated. It absorbs the material world. It may be that a "clash" occurs between the material and antimaterial worlds, as suggested by the scientists, and that the material worlds are destroyed, but there is no annihilation of the antimaterial world. *The eternally existing antimaterial world is unmanifested to the material scientist. He can simply have information of it insofar as the principles of its existence are contrary to the modes of the material world.* Full details of the antimaterial universe can be known only from the infallible source of liberated authorities who have thoroughly realized the constitution of the antimaterial principle.

This information is received by aural reception by a submissive disciple of the Personality of Godhead.

The Vedic knowledge was thus imparted into the heart of Brahmā, the first living being in the material creation. It was Brahmā who related this knowledge to the sage Nārada Muni. Similarly, the *Bhagavad-gītā* was spoken by the Personality of Godhead, Śrī Kṛṣṇa, to Vivasvān, the presiding deity of the sun, and when the aural chain of disciplic succession was broken, Lord Kṛṣṇa repeated the *Bhagavad-gītā* to Arjuna on the Battlefield of Kurukṣetra. At that time, Arjuna took the role of disciple and student in order to receive transcendental knowledge from Śrī Kṛṣṇa. In order to drive out all misgivings which the gross materialists of the world may have, Arjuna asked all relevant questions, and the answers were given by Kṛṣṇa so that any layman can understand them. Only those who are captivated by the glamour of the material world cannot accept the authority of Lord Śrī Kṛṣṇa. *One has to become thoroughly clean in habit and heart before one can understand the details of the antimaterial world. Bhakti-yoga* is a detailed scientific transcendental activity that both the neophyte and the perfect *yogī* can practice.

The material world is only a shadow representation of the antimaterial world, and intelligent men who are clean in heart and habit will be able to learn, in a nutshell, all the details of the antimaterial world from the text of the *Bhagavad-gītā*, and these are in actuality more exhaustive than material details. The basic details are as follows:

The presiding Deity of the antimaterial world is Śrī Kṛṣṇa, who exists in His original personality as well as in His many plenary expansions. This personality and His plenary expansions can be known only by antimaterial activities commonly known as *bhakti-yoga*, or devotional service. The Personality of Godhead is the supreme truth, and He is the whole antimaterial principle. The material principle as well as the antimaterial principle is an emanation from His person. He is the root of the complete tree. When water is poured onto the root of a tree, the branches and leaves are nourished automatically. And in the same way, when Śrī Kṛṣṇa, the Personality of Godhead, is worshiped, all details of the material worlds are enlightened, and the heart of the devotee is nourished without his having to work in a materialistic way. This is the secret of the *Bhagavad-gītā*.

The process of entering into the antimaterial world differs from materialistic processes. The individual living being can very easily enter the antimaterial world by practicing antimaterial activities while residing in the material world. But those who are truly gross materialists, who depend on the limited strength of experimental thought, mental speculation and materialistic science, find great difficulty in entering the antimaterial world. The gross materialist may try to approach the antimaterial worlds by endeavoring with spaceships, satellites, rockets, etc., which he throws into outer space, but by such means he cannot even approach the material planets in the higher regions of the material sky, and what to speak of those planets situated in the antimaterial sky, which is far beyond the material universe. Even the *yogīs* who have perfectly controlled mystic powers have great difficulty entering into that region. Master *yogīs* who control the antimaterial particle within the material body by practice of mystic powers can give up their material bodies at will at a certain opportune moment and can thus enter the antimaterial worlds through a specific thoroughfare which connects the material and antimaterial worlds. If they are at all able, they act in accordance with the prescribed method given in the *Bhagavad-gītā:*

"Those who have realized the Transcendence can

> reach the antimaterial world by leaving their material bodies during the period of *uttarāyaṇa*, that is, when the sun is on its northern path, or during auspicious moments in which the deities of fire and effulgence control the atmosphere."

The different deities, or powerful directing officers, are appointed to act in the administration of cosmic activities. Foolish people who are unable to see the intricacies of cosmic management laugh at the idea of personal management of fire, air, electricity, days, nights, etc., by demigods. But the perfect *yogīs* know how to satisfy these unseen administrators of material affairs and, taking advantage of the good will of these administrators, leave their material bodies at will during opportune moments arranged for entrance into the antimaterial universe or into the highest planets of the material sky. In the higher planets of the material world, the *yogīs* can enjoy more comfortable and more pleasant lives for hundreds of thousands of years, but life in those higher planets is not eternal. Those who desire eternal life enter into the antimaterial universe through mystic powers at certain opportune moments created by the demigod-administrators of cosmic affairs, administrators unseen by the gross materialists who reside on this seventh-class planet called "Earth."

> "Those who are not *yogīs* but who die at an opportune moment due to pious acts of sacrifice, charity, penance, etc., can rise to the higher planets after death, but are subject to return to this planet [Earth]. Their going forth takes place at a period known as *dhūma*; the dark, moonless half of the month; or when the sun is on its southern path."

In summary, the *Bhagavad-gītā* recommends that one adopt the means of devotional service, or antimaterial activities, if one wishes to enter the antimaterial world. Those who adopt the means of devotional service, as prescribed by the expert transcendentalist, are never disappointed in their attempts to enter the antimaterial world. Although the obstacles are many, the devotees of Lord Kṛṣṇa can easily overcome them by rigidly following the path outlined by the transcendental devotees. Such devotees, who are passengers progressing in the journey of life toward the antimaterial kingdom of God, are never bewildered. No one is cheated or disappointed when he adopts the guaranteed path of devotion for entrance into the antimaterial universe. One can easily attain all the results that are derived from the studies of the *Vedas*, performances of sacrifice, practices of penance and offerings of charity simply by the unilateral performance of devotional service, technically known as *bhakti-yoga*.

Bhakti-yoga is therefore the great panacea for all, and it has been made easy to practice, especially in this iron age, by Lord Kṛṣṇa Himself in His most sublime, liberal and munificent appearance as Lord Śrī Caitanya (1486–1534), who appeared in Bengal and spread the *saṅkīrtana* movement—singing, dancing, and chanting the names of God—throughout India. By Lord Caitanya's grace, one can quickly pick up the principles of *bhakti-yoga*. Thus all misgivings in the heart will disappear, the fire of material tribulation will be extinguished, and transcendental bliss will be ushered in.

In the fifth chapter of the *Brahma-saṁhitā* there is a description of the variegated planetary systems within the material world. It is also indicated in the *Bhagavad-gītā* that there are variegated planetary systems in hundreds of thousands of material universes, and that altogether these universes comprise only a fraction (one fourth) of the creative energy of the Godhead. The majority (three fourths) of the Lord's creative energy is manifested in the spiritual sky, called the *paravyoma* or the Vaikuṇṭhaloka. These instructions of the *Brahma-saṁhitā* and the *Bhagavad-gītā* may be finally confirmed by the material scientist as he researches into the existence of the antimaterial world.

In addition, a February 21, 1960, Moscow news release reported:

> "Russia's well-known professor of astronomy Boris Vorontsov-Veliaminov said that there must be an infinite number of planets in the universe inhabited by beings endowed with reason."

This statement of the Russian astronomer is a confirmation of the information given in the *Brahma-saṁhitā* (5.40), which states:

> *yasya prabhā prabhavato jagad-aṇḍa-koṭi-*
> *koṭiṣv aśeṣa-vasudhādi-vibhūti-bhinnam*
> *tad brahma niṣkalam anantam aśeṣa-bhūtaṁ*
> *govindam ādi-puruṣaṁ tam ahaṁ bhajāmi*

According to this quote from the *Brahma-saṁhitā*, there are not only infinite numbers of planets, as confirmed by the Russian astronomer, but there are also *infinite numbers of universes.* All these infinite universes with their infinite planets within are floating on and are produced from the Brahman effulgence emanating from the transcendental body of Mahā-Viṣṇu, who is worshiped by Brahmā, the presiding deity of the universe in which we are residing.

The Russian astronomer also confirms that all the planets—which are estimated to be not less than one hundred million—

are inhabited. In the *Brahma-saṁhitā* there is indication that in each and every one of the infinite number of universes there are infinite numbers of variegated planets.

The astronomer's view was seconded by Professor Vladimir Alpatov, a biologist, who maintained that some of the above-mentioned planets had reached a state of development corresponding to that of the earth. The report from Moscow continued:

> "It could be that life similar to that on Earth flourishes on such planets. Doctor of Chemistry Nikolai Zhirov, covering the problem of atmosphere on the planets, pointed out that the organism of a Martian, for instance, could very well adapt itself to normal existence with a low body temperature. He said that he felt that the gaseous composition of the atmosphere of Mars was quite suitable to sustain life of beings which have become adapted to it."

The adaptability of organisms in different varieties of planets is described in the *Brahma-saṁhitā* as *vibhūti-bhinnam,* i.e., each and every one of the innumerable planets within the universes is endowed with a particular type of atmosphere, and the living beings there are advanced in science, psychology, etc., according to the superiority or inferiority of the atmosphere. *Vibhūti* means "specific power," and *bhinnam* means "variegated." Scientists who are attempting to explore outer space in an attempt to reach other planets by mechanical means must realize that organisms adapted to the atmosphere of the earth cannot exist in the atmospheres of other planets. As such, man's attempts to reach the moon, the sun, or Mars will be completely futile because of the different atmospheres prevailing on those planets. Individually, however, one can attempt to go to any planet he desires, but this is only possible by psychological changes in the mind. Mind is the nucleus of the material body. The gradual evolutionary progress of the material body depends on psychological changes within the mind. The change of the bodily construction of a worm into that of a butterfly and, in modern medical science, the conversion of a man's body into that of a woman (or vice versa) are more or less dependent on psychological changes.

In the *Bhagavad-gītā* it is said that if a man, at the time of death, concentrates his mind upon the form of the Personality of Godhead, Śrī Kṛṣṇa, and while so doing relinquishes his body, he at once enters the spiritual existence of the antimaterial world. This means that anyone who trains the mind to turn from matter to the spiritual form of the Godhead by performance of the prescribed rules of devotional service can easily attain the kingdom of God, in the antimaterial sky. And of this there is no doubt.

And in the same way, if one desires to enter into any other planet of the material sky, he can go there just after quitting the present body (i.e., after death). Thus if someone wants to go to the moon, the sun or Mars, he can do so simply by performing acts for that purpose. The *Bhagavad-gītā* confirms this statement in the following words:

> "That upon which a person meditates at the time of death, quitting his body absorbed in the thought thereof—that particular thing he attains after death."

Mahārāja Bharata, despite a life of severe penances, thought of a stag at the time of his death and thus became a stag after death. However, he did retain a clear consciousness of his past life and realized his mistake. *It is important to realize that one's thoughts at the time of death are influenced by the actual deeds which one performs during his life.*

In the *Śrīmad-Bhāgavatam* (Third Canto, chapter thirty-two), the process of entering the moon is described as follows:

> "Materialistic-minded men, who have no information of the kingdom of God, are always mad after material acquisition of wealth, fame and adoration. Such men are interested in the progressive weal of their particular family unit for their own self-satisfaction and so are also interested in the progress of social and national welfare. These men attain their desired objects by material activities. They are mechanically engaged in the ritualistic discharge of prescribed duties and are consequently inclined to satisfy the Pitās, or bygone forefathers, and controlling demigods by performance of sacrifices as prescribed by the revealed scriptures. Addicted to such acts of sacrifices and ceremonial observances, such souls enter into the moon after death. When one is thus promoted to the moon, he receives the capacity to enjoy the drinking of *soma-rasa*, a celestial beverage. The moon is a place where the demigod Candra is the predominating deity. The atmosphere and amenities of life there are far more comfortable and advantageous than those here on earth. After reaching the moon, if a soul does not utilize the opportunity for promotion to better planets, he is degraded and forced to return to earth or a similar planet. However, materialistic persons, although they may attain to the topmost planetary system, are certainly annihilated at the time of the cosmic manifestation's dissolution."

As far as the planetary system of the spiritual sky is concerned, there are unlimited Vaikuṇṭha planets in the *para-*

vyoma. The Vaikuṇṭhas are spiritual planets which are manifestations of the internal potency of the Lord, and the ratio of these planets to the material planets (external energy) in the material sky is three to one. So the poor materialist is busy making political adjustments on a planet which is most insignificant in God's creation. To say nothing of this planet earth, the whole universe with innumerable planets throughout the galaxies is comparable to a mustard seed in a bag full of mustard seeds. But the poor materialist makes plans to live comfortably here and thus wastes his valuable human energy in something which is doomed to frustration. Instead of wasting his time with this plan-making business, he might have sought the life of plain living and high spiritual thinking and thus saved himself from perpetual materialistic unrest.

Even if a materialist wants to enjoy developed material facilities, he can transfer himself to planets where he can experience material pleasures much more advanced than those available on the earth planet. But the best plan is to prepare oneself to return to the spiritual sky after leaving the body. However, if one is intent on enjoying material facilities, one can transfer himself to other planets in the material sky by utilizing yogic powers. The playful spaceships of the astronauts are but childish entertainments and are of no use for this purpose.

The *aṣṭāṅga-yoga* system is also materialistic, inasmuch as it teaches one to control the movements of air within the material body. The spiritual spark, the soul, is floating on air within the body, and inhalation and exhalation are the waves of that air containing the soul. Therefore the *yoga* system is a materialistic art of controlling this air by transferring it from the stomach to the navel, from the chest to the collarbone and from there to the eyeballs and from there to the cerebellum and from there to any desired planet. The velocities of air and light are taken into consideration by the material scientist, but he has no information of the velocity of the mind and intelligence. We have some limited experience of the velocity of the mind, because in a moment we can transfer our minds to places hundreds of thousands of miles away. Intelligence is even finer. Finer than intelligence is the soul, which is not matter like mind and intelligence but is spirit, or antimatter. The soul is hundreds of thousands of times finer and more powerful than intelligence. We can thus only imagine the velocity of the soul in its traveling from one planet to another. Needless to say, the soul travels by its own strength and not with the help of any kind of material vehicle.

The bestial civilization of eating, sleeping, fearing and sense-gratifying has misled modern man into forgetting how powerful a soul he is. As we have already described, the soul is a spiritual spark which is many, many times more illuminating, dazzling and powerful than the sun, the moon or electricity. Human life is spoiled when man does not realize his real identity with his soul. Lord Caitanya appeared with Lord Nityānanda to save man from this type of misleading civilization.

Śrīmad-Bhāgavatam also describes how *yogīs* can travel to all the planets in the universe. When the vital force is lifted to the cerebellum, there is every chance of this force bursting out from the eyes, nose, ears, etc., as these are places which are known as the seventh orbit of the vital force. But the *yogīs* can block these holes by complete suspension of air. The *yogī* then concentrates the vital force in the middle position, that is, between the eyebrows. At this position, the *yogī* can think of the planet into which he wants to enter after leaving the body. He can then decide whether he wants to go to the abode of Kṛṣṇa in the transcendental Vaikuṇṭhas,. from which he will not be required to descend into the material world, or to travel to higher planets in the material universe. The perfect *yogī* is at liberty to do either.

For the perfect *yogī* who has attained success in the method of leaving his body in perfect consciousness, transferring from one planet to another is as easy as an ordinary man's walking to the grocery store. As already discussed, the material body is just a covering of the spiritual soul. Mind and intelligence are the undercoverings, and the gross body of earth, water, air, etc., is the overcoating of the soul. As such, any advanced soul who has realized himself by the yogic process, who knows the relationship between matter and spirit, can leave the gross dress of the soul in perfect order and as he desires. By the grace of God, we have complete freedom. Because the Lord is kind to us, we can live anywhere—either in the spiritual sky or in the material sky, upon whichever planet we desire. However, misuse of this freedom causes one to fall down into the material world and suffer the threefold miseries of conditioned life. The living of a miserable life in the material world by dint of the soul's choice is nicely illustrated by Milton in *Paradise Lost*. Similarly, by choice the soul can regain paradise and return home, back to Godhead.

At the critical time of death, one can place the vital force between the two eyebrows and decide where he wants to go. If he is reluctant to maintain any connection with the material world, he can, in less than a second, reach the transcendental Vaikuṇṭha world and appear there in a completely spiritual body just suitable for him in the spiritual atmosphere. He has simply to desire to leave the material world in both finer and grosser forms and then move the vital force to the topmost part of the skull and leave the body from the hole in the skull called the *brahma-randhra*. This is the highest perfection in the practice of *yoga*.

Of course, man is endowed with free will, and as such if he does not want to free himself of the material world he may enjoy the life of *brahma-pāda* (occupation of the post of Brahmā) and visit Siddhaloka, the planets of materially perfect beings who have full abilities to control gravity, space, time, etc. To visit these higher planets in the material universe, one need not give up his mind and intelligence (finer matter), but need only give up his body made of grosser matter.

Man-made satellites and mechanical space vehicles will never be able to carry human beings to the planets of outer space. Men cannot even go on their much-advertised trips to the moon, for, as we have already stated, the atmosphere on such higher planets is different from the atmosphere here on earth. Each and every planet has its particular atmosphere, and if one wants to travel to any particular planet within the material universe, one has to have a material body exactly adapted to the climatic condition of that planet. For instance, if one wants to go from India to Europe, where the climatic condition is different, one has to change his dress accordingly. Similarly, a complete change of body is necessary if one wants to go to the transcendental planets of Vaikuṇṭha.

If one wants to go to the higher material planets, he can keep his finer dress of mind, intelligence and ego, but he has to leave his gross dress (body) made of earth, water, fire, etc. When one goes to a transcendental planet, however, it is necessary to change both the finer and gross bodies, *for one has to reach the spiritual sky completely in a spiritual form.* This change of dress will take place automatically at the time of death if one so desires. But this desire is possible at death only if the desire is cultivated during life. When one performs fruitive actions, one cultivates desires in relation with the material world, but when one practices devotional service, one cultivates a desire for the kingdom of God. The following details outline a general practice by which one can *prepare himself for an easy journey to the Vaikuṇṭha (antimaterial) planets, where life is free from birth, old age, disease and death.*

General practice (positive functions):

1. The serious candidate must accept a bona fide spiritual master in order to be trained scientifically. Because the senses are material, it is not at all possible to realize the Transcendence by them. Therefore the senses have to be spiritualized by the prescribed method under the direction of the spiritual master.
2. When the student has chosen a bona fide spiritual master, he must take proper initiation from him. This marks the beginning of spiritual training.
3. The candidate must be prepared to satisfy the spiritual master in every way. A bona fide spiritual master who is fully cognizant of the methods of spiritual science, learned in the spiritual scriptures such as the *Bhagavad-gītā*, *Vedānta-sūtras*, *Śrīmad-Bhāgavatam* and *Upaniṣads*, and who is also a realized soul who has made a tangible connection with the Supreme Lord, is the transparent medium by which the willing candidate is led to the path of the Vaikuṇṭhas. The spiritual master must be satisfied in all respects, because simply by his good wishes a candidate can make wonderful progress along the path.
4. The intelligent candidate places intelligent questions to the spiritual master in order to clear his path of all uncertainties. The spiritual master shows the way, not whimsically but in accordance with the principles of the authorities who have actually traversed the path. The names of these authorities are disclosed in the scriptures, and one has simply to follow them under the direction of the spiritual master. The spiritual master never deviates from the path of the authorities.
5. The candidate should always try to follow in the foot-

steps of the great sages who have practiced the method and obtained success. This should be taken as a motto in life. One should not superficially imitate them, but should follow them sincerely in terms of the particular time and circumstances.

6. The candidate must be prepared to change his habits in terms of the instructions contained in the books of authority, and for the satisfaction of the Lord he must be prepared to sacrifice both sense gratification and sense abnegation, following the example of Arjuna.
7. The candidate should live in a spiritual atmosphere.
8. He must be satisfied with as much wealth as is sufficient for maintenance only. He should not try to amass more wealth than is necessary to sustain himself in a simple way.
9. He must observe the fasting dates, such as the eleventh day of the growing and waning moon.
10. He must show respect to the banyan tree, the cow, the learned *brāhmaṇa* and the devotee.

These are the first steppingstones on the path of devotional service. Gradually one has to adopt other items, which are negative in character:

11. One should avoid offenses in the discharge of devotional service and in chanting the holy names.
12. He should avoid extensive association with nondevotees.
13. He must not take on unlimited disciples. This means that a candidate who has successfully followed the first twelve items can also become a spiritual master himself, just as a student becomes a monitor in class with a limited number of disciples.

14. He must not pose himself as a vastly learned man simply by quoting statements in books. He must have solid knowledge of the necessary books without superfluous knowledge in others.
15. A regular and successful practice of the above fourteen items will enable the candidate to maintain mental equilibrium even amidst great trials of material loss and gain.
16. In the next stage, the candidate does not become afflicted by lamentation and illusion.
17. He does not deride another's mode of religion or worship, nor does he deride the Personality of Godhead or His devotees.
18. He never tolerates blasphemy against the Lord or His devotees.
19. He should not indulge in the discussion of topics dealing with the relationship between man and woman; nor should he engage in useless topics concerning others' family affairs.
20. He should not inflict pain—either in body or in mind—upon other living beings, whomsoever they may be.

Out of the above twenty items, the first three positive items are imperative and most essential for the serious candidate.

There are forty-four other items to be followed by the serious candidate, but Lord Caitanya has selected five as the most important. These were selected owing to the present conditions of civic life. They are as follows:

1. *One should associate with the devotees.* Association with devotees is made possible by hearing them attentively, by asking them relevant questions, by supplying them

food and accepting food from them, and by giving them charity and accepting from them whatever they offer.

2. *One should chant the holy name of the Lord in all circumstances.* The chanting of the Lord's name is an easy and inexpensive process of realization. One can chant any of the innumerable names of the Lord at any time. One should try to avoid offenses. There are ten offenses which one can commit while chanting the transcendental names, and these should be avoided as far as possible, but in any event, one should try to chant the holy names of the Lord at all times.
3. *One should hear the transcendental topics enunciated in the* Śrīmad-Bhāgavatam. This hearing is made possible through platform lectures by bona fide devotees and by authorized translations of the *Bhāgavatam.*
4. *One should make his home at Mathurā, the birthplace of Lord Kṛṣṇa.* Or one may make his home as good as Mathurā by installing the Deity of the Lord to be worshiped by all members of the family after proper initiation from the spiritual master.
5. *One should worship the installed Deity with attention and devotion so that the whole atmosphere of one's home becomes a replica of the Lord's abode.* This is made possible by the direction of the spiritual master, who knows the transcendental art and can show the candidate the proper method.

The above five items can be adopted by any man in any part of the world. Thus anyone can prepare himself for returning home, back to Godhead, by the simple method recognized by authorities such as Lord Śrī Caitanya Mahāprabhu, who specifically advented Himself to deliver the fallen souls of this age.

For further details on this subject, one should read literatures like the *Bhakti-rasāmṛta-sindhu,* of which we have presented an English summary study entitled *The Nectar of Devotion.*

The whole process of transferring oneself to the spiritual sky involves gradually liquidating the material composition of the gross and subtle coverings of the spirit soul. The above-mentioned five items of devotional activities are so spiritually powerful that their performance by a devotee, even in the preliminary stage, can very quickly promote the sincere executor to the stage of *bhāva* (the stage just prior to love of Godhead), or emotion on the spiritual plane, which is transcendental to mental and intellectual functions. And complete absorption in love of God makes one fit to be transferred to the spiritual sky just after leaving the material tabernacle. The perfection of love of God by a devotee actually situates him on the spiritual platform, even though he may still maintain a gross material body. He becomes like an iron rod which, when in contact with fire, actually ceases to be iron and acts like fire. These things are made possible by the Lord's inscrutable and inconceivable energy, which material science has not the scope to calculate. One should therefore engage himself in devotional service with absolute faith, and to make his faith steadfast one should seek the association of the Lord's standard devotees in person (if possible) or by thinking of them. This association will help one develop factual devotional service to the Lord, which will cause all material misgivings to disappear like a flash of lightning. All these different stages of spiritual realization will be personally felt by the candidate, and this will create in him a firm belief that he is making positive progress on the way to the spiritual sky. Then he will become sincerely attached to the Lord and His abode. Such is the gradual process of evolving love of God, which is the prime necessity for the human form of life.

His Divine Grace
A. C. Bhaktivedanta Swami Prabhupāda
Founder-*ācārya* of the International Society for Krishna Consciousness

Lord Kṛṣṇa descends to this material world from the antimaterial world and speaks the *Bhagavad-gītā* to His friend and disciple Arjuna, thus giving complete information of His superior energy. (p. 3)

In Bengal, in 1486, Lord Kṛṣṇa appeared as Lord Śrī Caitanya Mahāprabhu, His most liberal and munificent form, to spread the *saṅkīrtana* movement of singing, dancing, and chanting the names of God. (p. 19)

The infinite universes with their infinite planets are produced from the Brahman effulgence emanating from the transcendental body of Mahā-Viṣṇu, an expansion of Lord Kṛṣṇa. (p. 20)

The Supreme Being, who is the supreme intelligence, is the ultimate creator, the all-attractive Personality of Godhead, Śrī Kṛṣṇa. (p. 39)

The ultimate goal of all living entities is to return home, back to Godhead, to enjoy an eternal life of bliss and knowledge by engaging in loving devotional service to Lord Śrī Kṛṣṇa. (p. 43)

The spiritual master is the external manifestation of God, who is situated in everyone's heart as the Supersoul. (p. 52)

It is not advancement of knowledge to go to the moon after ten years of effort and take a rock and come back. (p. 55)

There are instances in history of great personalities, including sages and kings, who attained perfection by this process. Some of them attained success even by adhering to *one single item* of devotional service with faith and perseverance. Some of these personalities are listed below.

1. Emperor Parīkṣit attained the spiritual platform simply by *hearing* from such an authority as Śrī Śukadeva Gosvāmī.
2. Śrī Śukadeva Gosvāmī attained the same simply by *recitation*, verbatim, of the transcendental message which he received from his great father, Śrī Vyāsadeva.
3. Emperor Prahlāda attained spiritual success by *remembering* the Lord constantly, in pursuance of instructions given by Śrī Nārada Muni, the great saint and devotee.
4. Lakṣmījī, the goddess of fortune, attained success simply by sitting and *serving* the lotus feet of the Lord.
5. King Pṛthu attained success simply by *worshiping* the Lord.
6. Akrūra, the charioteer, attained success simply by *chanting prayers* for the Lord.
7. Hanumān (Mahāvīra), the famous nonhuman devotee of Lord Śrī Rāmacandra, attained success simply by *carrying out the orders* of the Lord.
8. Arjuna, the great warrior, attained the same perfection simply by *making friends with the Lord*, who delivered the message of the *Bhagavad-gītā* to enlighten Arjuna and his followers.
9. Emperor Bali attained success by *surrendering everything* unto the Lord, including his personal body.

These are nine standard modes of devotional service to the

Lord, and a candidate can choose to adopt any one, two, three, four or all, however he likes. All the services rendered to the Absolute are in themselves absolute, with none of the quantitative or qualitative differences found on the material platform. On the spiritual platform everything is identical with everything else, although there is transcendental variegatedness. Emperor Ambarīṣa adopted all the above nine items, and he attained perfect success. He engaged his mind on the lotus feet of the Lord, his voice in describing the spiritual world, his hands in cleansing the temple of the Lord, his ears in submissively hearing the words of Lord Śrī Kṛṣṇa, his eyes in viewing the Deities of the Lord, his body in touching the bodies of the devotees, his nostrils in smelling the flowers offered to the Lord, his tongue in tasting the food offered to the Lord, his legs in visiting the temple of the Lord, and all the energy of his life in executing the services of the Lord, without in the least desiring his own sense gratification. All these activities helped him attain the perfect stage of life, which defeats all dexterities of material science.

It is therefore incumbent upon all human beings to adopt these principles of spiritual realization for the perfection of life. A human being's only obligation is spiritual realization. Unfortunately, in modern civilization, human society is too busy discharging national duties. Actually, national duties, social duties and humanitarian duties are obligatory only for those who are bereft of spiritual duties. As soon as a man takes his birth on this earth, not only does he have national, social and humanitarian obligations, but he also has obligations to the demigods who supply air, light, water, etc. He also has obligations to the great sages who have left behind them vast treasure-houses of knowledge to guide him through life, and also to all kinds of living beings, to his forefathers, to his family members and so on and so forth. But as soon as one engages himself in the one single obligatory duty—the duty of spiritual perfection—then he automatically liquidates all other obligations without having to make separate efforts.

A devotee of the Lord is never a disturbing element in society—on the contrary, he is a great social asset. Since no sincere devotee is attracted to sinful actions, as soon as a man becomes a pure devotee he can do inestimable selfless service to society for the peace and prosperity of all concerned, in this life and in the next. But even if such a devotee commits some offense, the Lord Himself rectifies it in no time. Therefore, there is no need for a devotee to cultivate materialistic knowledge, nor does a devotee need to artificially renounce everything and live as a hermit. He can simply remain at home and execute devotional service smoothly in any order of life. And there are instances in history of extremely cruel men becoming kindhearted simply by the execution of devotional service. Knowledge and abnegation follow automatically in the life of a pure devotee without his having to make extraneous effort.

This spiritual art and science of devotional service is the highest contribution of Indian sages to the rest of the world. Therefore everyone who has taken his birth in India has an obligation to perfect his life by adopting the principles of this great art and science and distributing it to the rest of the world, which is still ignorant of the ultimate aim of life. Human society is destined to reach this stage of perfection by gradual development of knowledge. Indian sages, however, have already reached that position. Why do others have to wait for thousands and thousands of years to attain their heights? Why not give them the information immediately in a systematic way, so that they may save time and energy? They should take advantage of human life, which they may have labored millions of years to attain.

A Russian fiction writer is now contributing suggestions to the rest of the world that scientific progress can help man to live forever. Of course, he does not believe in a Supreme Being who is the creator. Yet we welcome his suggestion because we know that actual progress in scientific knowledge will certainly take men to the spiritual sky and inform the scientists that there is a supreme creator who has full potencies beyond all materialistic scientific conceptions.

As mentioned, every living being is eternal but has to change his outer coverings, gross and subtle, and this changing process is technically known as life and death. As long as a living being has to put on the shackles of material bondage, there is no relief from this changing process, which continues even in the highest stage of material life. The Russian fiction writer may speculate, as fiction writers are apt to do, but saner people with some knowledge of natural law will not agree that man can live forever within this material world.

A naturalist can see the general course of material nature simply by studying a fruit in a tree. A small fruit develops from a flower, grows, stays for some time on a branch, becomes full-grown, ripens, then begins to dwindle daily until it finally falls from the tree and commences to decompose into the earth and at last mingles with the earth, leaving behind its seed, which in its turn grows to become a tree and produces many fruits in time, which will all meet the same fate, and so on and so on.

Similarly, a living being (as a spiritual spark, a part of the Supreme Being) takes its organic form in the womb of a mother just after sexual intercourse. It grows little by little within the womb, is born, then continues growing, becomes a child, a boy, a youth, an adult and an old man, and then finally dwindles and meets death, despite all the good wishes and hopeful pipe dreams of fiction writers. By comparison, there is no difference between the man and the fruit. Like the fruit, the man may leave behind him his seeds of numerous children, but he cannot exist eternally within his material body, due to the law of material nature.

How can anyone ignore the law of material nature? No material scientist can change the stringent laws of nature, however boastful he may be. No astronomer can change the course of the planets—he can only manufacture a paltry toy planet, which he calls a satellite. Foolish children may be impressed by this and may give a great deal of credit to the inventors of modern satellites, sputniks, etc., but the saner section of humanity gives more credit to the creator of the gigantic satellites, namely the sun, stars and planets, of which the material scientist can see no end. If a small toy satellite has a creator in Russia or America, it is reasonable that the gigantic satellites have their creator in the spiritual sky. If a toy satellite requires so many scientific brains for its manufacture and its orbiting, what kind of subtle and perfect brain created galaxies of stars and maintains them in their orbits? Thus far the atheistic class have not been able to answer this.

Nonbelievers put forward their own theories of the creation, which usually result in statements such as, "It's hard to understand," "Our imagination cannot conceive it, but it's quite possible," "It's incomprehensible," and so forth. This only means that their information has no authoritative basis and is not backed by scientific data. They simply speculate. However, authorized information is available in the *Bhagavad-gītā*. For instance, the *Bhagavad-gītā* informs us that within the material world there are living beings whose duration of life covers 4,300,000 x 1,000 x 2 x 30 x 12 x 100 solar years. We accept the *Bhagavad-gītā* as authority because this book of knowledge was so accepted by India's great sages like Śaṅkarācārya, Śrī Rāmānujācārya, Śrī Madhvācārya and Śrī Caitanya Mahāprabhu. The *Bhagavad-gītā* indicates that in the material world all living forms are subject to decay and death, regardless of their duration of life.

Therefore all material shapes are subject to the law of change, although the material energy itself is conserved. As energy, everything is eternal, but in the material world matter takes shape, remains for some time, develops into maturity, grows old, begins to dwindle and at last disappears again. This is the case with all material objects. The materialist's suggestion that beyond the material sky there is "some other form" that is outside the boundary of visibility and that is strange and inconceivable is but a faint indication of the spiritual sky. However, the basic principle of spirit is much closer, for it functions within all living beings. When that spiritual principle is out of the material body, the material body has no life. Within the body of a child, for instance, the spiritual principle is present, and therefore changes take place in the body and it develops. But if the spirit leaves the body, the development stops. This law is applicable to every material object. Matter transforms from one shape to another when it is in contact with spirit. Without spirit there is no transformation. The entire universe develops in that way. It emanates from the energy of the Transcendence because of the spiritual force which is His, and it develops into gigantic forms like the sun, moon and earth There are fourteen divisions of planetary systems, and although they are all different in dimension and quality, the same principle of development holds true for all. The spiritual force is the creator, and only by this spiritual principle do transformation, transition and development take place.

Life is definitely not generated simply by a chemical combination or any other material reaction, as many foolish men claim. Material interaction is set in motion by the Supreme Spiritual Being, who creates a favorable circumstance to accommodate the spiritual living force. The Supreme Spiritual

Being handles matter in an appropriate way, as determined by His free will. For example, building materials do not "automatically react" and suddenly assume the shape of a residential house. A living spiritual being handles matter appropriately by his free will and thus constructs the house. Similarly, matter is the ingredient only, but the Supreme Spirit is the creator. Only a man with a poor fund of knowledge avoids this conclusion. The creator may remain unseen in the background, but that does not mean there is no creator. One should not be illusioned simply by the gigantic form of the material universe. Rather, one should learn to discern the existence of a supreme intelligence behind all these material manifestations. The Supreme Being, who is the supreme intelligence, *is the ultimate creator, the all-attractive Personality of Godhead, Śrī Kṛṣṇa.* Although one may not be aware of this, there is definite information of the creator given in Vedic literatures such as the *Bhagavad-gītā* and especially the *Śrīmad-Bhāgavatam.*

When a satellite is thrown into outer space, a child may not understand that there are scientific brains behind it, but an intelligent adult realizes that scientific brains on earth are controlling the satellite. Similarly, less intelligent persons do not have information of the creator and His eternal abode in the spiritual world, which is far beyond our range of visibility, but in actuality there is a spiritual sky with spiritual planets that are more spacious and greater in number than the planets in the material sky. From the *Bhagavad-gītā* we receive information that the material universe constitutes only a fraction (one fourth) of the creation. Such information is extensively available in the *Śrīmad-Bhāgavatam* and other Vedic literatures.

If living energy could be generated in the scientist's laboratory by "the interaction of certain physical and chemical combinations," then why haven't the boastful material scientists been able to manufacture life? They should know definitely that the spiritual force is distinct from matter and that such energy is not possible to produce by any amount of material adjustment. At present Russians and Americans are undoubtedly very much advanced in many departments of technological science, but they are still ignorant of the spiritual science. They will have to learn from superior intelligence in order to make a perfect and progressive human society.

The Russians are unaware that in the *Śrīmad-Bhāgavatam* the socialist philosophy is most perfectly described. The *Bhāgavatam* instructs that whatever varieties of wealth exist—such as natural resources (agricultural, mining, etc.)—all are created by the ultimate creator, and therefore every living being has a right to take his share of them. It is further said that a man should possess only as much wealth as is sufficient to maintain his body, and that if he desires more than that, or if he takes more than his share, he is subject to punishment. It is also stated that animals should be treated as one's own children.

We believe that no nation on earth can describe socialism as well as the *Śrīmad-Bhāgavatam.* Living beings other than humans can be treated as brothers and children only when one has a full conception of the creator and the actual constitution of the living being.

Man's desire to be deathless is realized only in the spiritual world. As stated at the beginning of this essay, a desire for eternal life is a sign of dormant spiritual life. The aim of human civilization should be targeted to that end. It is possible for every human being to transfer himself to that spiritual realm by the process of *bhakti-yoga*, as described herein. It is a great science, and India has produced many scientific literatures by which the perfection of life may be realized.

Bhakti-yoga is the eternal religion of man. At a time when material science predominates all subjects—including the tenets of religion—it would be enlivening to see the principles of the eternal religion of man from the viewpoint of the modern scientist. Even Dr. S. Radhakrishnan admitted at a world religion conference that religion will not be accepted in modern civilization if it is not accepted from a scientific point of view. In reply, we are glad to announce to the lovers of the truth that *bhakti-yoga* is the eternal religion of the world and is intended for all living beings, who are all eternally related with the Supreme Lord.

Śrīpāda Rāmānujācārya defines the word *sanātana,* or "eternal," as that which has neither beginning nor end. When we speak of *sanātana-dharma,* eternal religion, we take this definition for granted. That which has neither beginning nor end is unlike anything sectarian, which has limits and boundaries. In the light of modern science it will be possible for us to see *sanātana-dharma* as the main occupation of all the people of the world—nay, of all living entities of the universe. Non-*sanātana* religious faith has some beginning in the annals of man, but there is no historic origin of *sanātana-dharma* because it eternally remains with the living entities.

When a man professes to belong to a particular faith—Hindu, Muslim, Christian, Buddhist or any other sect—because of the particular time and circumstances of his birth, such designations are called non-*sanātana-dharma.* A Hindu may become a Muslim or a Muslim may become a Hindu or Christian, etc., but in all circumstances there is one constant. In all circumstances, he is rendering service to others. A Hindu, Muslim, Buddhist or Christian is in all circumstances a servant of someone. The particular type of faith professed is not *sanātana-dharma. Sanātana-dharma* is the constant companion of the living being, the unifier of all religions. *Sanātana-dharma* is the rendering of service.

In the *Bhagavad-gītā* there are several references to that which is *sanātana*. Let us learn the import of *sanātana-dharma* from this authority.

There is reference to the word *sanātanam* in the tenth verse of the seventh chapter, in which the Lord says that He is the eternal fountainhead of everything and is therefore *sanātanam*. The fountainhead of everything is described in the *Upaniṣads* as the complete whole. All emanations of the fountainhead are also complete in themselves, but although many complete units emanate from the complete *sanātana* fountainhead, the *sanātana* head does not diminish in quality or quantity.* That is because the nature of *sanātana* is unchangeable. Anything that changes under the influence of time and circumstances is not *sanātana*. Therefore anything that changes whatsoever in form or quality cannot be accepted as *sanātana*. To give a material example, the sun has been disseminating its rays for hundreds of millions of years, and yet although it is a materially created object, its form and rays are still unchanged. Therefore, that which is never created cannot change in form or quality, even though He is the seedling source of everything.

The Lord claims to be *the father of all species of life*. He claims that all living beings—regardless of what they are—are part and parcel of Him. Consequently, the *Bhagavad-gītā* is meant for all of them. In the Gītā there is information of this *sanātana* nature of the Supreme Lord. There is also information of His abode, which is far beyond the material sky, and of the *sanātana* nature of the living beings.

Lord Kṛṣṇa, in the *Bhagavad-gītā*, also informs us that this material world is full of miseries in the shape of birth, old age, disease and death. Even in the topmost planet of the material universe, Brahmaloka, these miseries are present. Only in His

---

*See *Śrī Īśopaniṣad*, Invocation.

own abode is there a total absence of misery. In that abode there is no need of light from sun, moon or fire because the planets there are self-luminous. Life there is perpetual and full of knowledge and bliss. That is what is known as the *sanātana-dhāma*. It is therefore natural to conclude that the living entities must return home, back to Godhead, to enjoy life in the *sanātana-dhāma* with the *sanātana-puruṣa*, or the *puruṣottama*, Lord Śrī Kṛṣṇa. They must not remain to rot in this miserable land of material existence. There is no happiness in the material sphere—even in Brahmaloka—so plans and activities for elevation to higher planets within the material universe are carried out by those who are less intelligent. Less intelligent men also take shelter of demigods and only derive benefits which endure for a limited period. Thus their religious principles and the benefits derived therefrom are only temporary. The intelligent man, however, abandons all such so-called religious engagements and takes shelter of the Supreme Personality of Godhead and thus receives absolute protection from the Almighty Father. *Sanātana-dharma* is therefore the process of *bhakti-yoga*, by which one can come to know the *sanātana* Lord and His *sanātana* abode. By this process only can one return to the spiritual universe, the *sanātana-dhāma*, to take part in the *sanātana* enjoyment prevailing there.

Those who are followers of *sanātana-dharma* may henceforward take up those principles in the spirit of the *Bhagavad-gītā*. There is nothing barring anyone from adopting the eternal principles. Even persons who are less enlightened can return to Godhead. This is the version taught by *Śrīmad-Bhāgavatam* and by the Supreme Lord Himself in the *Bhagavad-gītā*. Mankind should be given a chance to take advantage of this opportunity. Because the *Bhagavad-gītā* was spoken in the land of Bhārata-varṣa, every Indian has the responsibility to broadcast the message of real *sanātana-dharma* in the other parts of the world. Especially at the present moment, misguided men are suffering in the darkness of materialism, and their so-called learning has enabled them to discover the atomic bomb. They are consequently on the verge of annihilation. *Sanātana-dharma*, however, will teach them about the real purpose of life, and they will benefit by its propagation.

2

Varieties of Planetary Systems

In these days, when men are trying to go to the moon, people should not think that Kṛṣṇa consciousness is concerned with something old-fashioned. When the world is progressing to reach the moon, we are chanting Hare Kṛṣṇa. But people should not misunderstand and assume that we are lagging behind modern scientific advancement. We have already passed all scientific advancement. In the *Bhagavad-gītā* it is said that man's attempt to reach higher planets is not new. Newspaper headlines read "Man's First Steps on the Moon," but the reporters do not know that millions and millions of men went there and came back. This is not the first time. This is an ancient practice. In the *Bhagavad-gītā* (8.16) it is clearly stated, *ābrahma-bhuvanāl lokāḥ punar āvartino 'rjuna:* "My dear Arjuna, even if you go to the highest planetary system, which is called Brahmaloka, you will have to come back." Therefore, interplanetary travel is not new. It is known to the Kṛṣṇa conscious devotees.

Since we are Kṛṣṇa conscious, we take what Kṛṣṇa says to be

the Absolute Truth. According to the Vedic literature, there are many planetary systems. The planetary system in which we are living is called Bhūrloka. Above this planetary system is Bhuvarloka. Above that is Svarloka (the moon belongs to the Svarloka planetary system). Above Svarloka is Maharloka; above that is Janaloka; and above that is Satyaloka. Similarly, there are lower planetary systems. Thus there are fourteen planetary systems within this universe, and the sun is the chief planet. The sun is described in the *Brahma-saṁhitā* (5.52):

yac-cakṣur eṣa savitā sakala-grahāṇāṁ
rājā samasta-sura-mūrtir aśeṣa-tejāḥ
yasyājñayā bhramati saṁbhṛta-kāla-cakro
govindam ādi-puruṣaṁ tam ahaṁ bhajāmi

"I worship Govinda [Kṛṣṇa], the primeval Lord, by whose order the sun assumes immense power and heat and traverses its orbit. The sun, which is the chief among all planetary systems, is the eye of the Supreme Lord." Actually, without the sun we cannot see. We may be very proud of our eyes, but we cannot even see our next-door neighbor. People challenge, "Can you show me God?" But what can they see? What is the value of their eyes? God is not cheap. We cannot see anything, not to speak of God, without sunshine. Without sunlight we are blind. At night, we cannot see anything, and therefore we use electricity because the sun is not present.

It is not that there is only one sun in the cosmic manifestation; there are millions and trillions of suns. That is also stated in the *Brahma-saṁhitā* (5.40):

yasya prabhā prabhavato jagad-aṇḍa-koṭi-
koṭiṣv aśeṣa-vasudhādi-vibhūti-bhinnam
tad brahma niṣkalam anantam aśeṣa-bhūtaṁ
govindam ādi-puruṣaṁ tam ahaṁ bhajāmi

The spiritual bodily effulgence of the Supreme Personality of Godhead, Kṛṣṇa, is called the *brahma-jyoti*, and in that *brahma-jyoti* there are countless planets. Just as within the sunshine there are innumerable planets, in the shining effulgence of the body of Kṛṣṇa there are innumerable planets and universes. We have knowledge of many universes, and in each universe there is a sun. Thus there are millions and billions of universes and millions and billions of suns and moons and planets. But Kṛṣṇa says that if one tries to go to any one of these planets, he will simply waste his time.

Now someone has gone to the moon, but what will human society gain from it? If, after spending so much money, so much energy and ten years of effort, one goes to the moon and simply touches it, what is the benefit of that? Can one remain there and call his friends to come? And even if one goes there and remains, what will be the benefit? As long as we are in this material world, either on this planet or other planets, the same miseries—birth, death, old age and disease—will follow us. We cannot rid ourselves of them.

If we go to live on the moon—assuming it is possible—even with an oxygen mask, how long could we stay? Furthermore, even if we had the opportunity to stay there, what would we gain? We might gain a little longer life perhaps, but we could not live there forever. That is impossible. And what would we gain by a longer life? *Taravaḥ kiṁ na jīvanti:* are not the trees living for many, many years? Near San Francisco I have seen a forest where there is a tree seven thousand years old. But what is the benefit? If one is proud of standing in one place for seven thousand years, that is not a very great credit.

How one goes to the moon, how he comes back, etc., is a great story, and this is all described in the Vedic literature. It is not a very new process. But the aim of our Kṛṣṇa consciousness society is different. We are not going to waste our valuable time. Kṛṣṇa says, "Don't waste your time attempting to go to this planet or to that planet. What will you gain? Your material miseries will follow you wherever you go." Therefore, in the *Caitanya-caritāmṛta* (*Ādi* 3.97) it is very nicely said by the author:

keha pāpe, keha puṇye kare viṣaya-bhoga
bhakti-gandha nāhi, yāte yāya bhava-roga

"In this material world someone is enjoying and someone is not enjoying, but actually everyone is suffering, although some people think that they are enjoying, whereas others realize that they are suffering." Actually everyone is suffering. Who in this material world does not suffer disease? Who does not suffer from old age? Who does not die? No one wants to grow old or suffer from disease, but everyone must do so. Where then is the enjoyment? This enjoyment is all nonsense because within this material world there is no enjoyment. It is simply our imagination. One should not think, "This is enjoyment, and this is suffering." Everything is suffering! Therefore, it is stated in the *Caitanya-caritāmṛta*, "The principles of eating, sleeping, mating and defending will always exist, but they will exist in different standards." For example, the Americans have taken birth in America as a result of pious activities performed in previous lifetimes. In India the people are poverty-stricken and are suffering, but although the Americans are eating very nicely buttered bread and the Indians are eating without butter, they are both eating nevertheless. The fact that India is poverty-stricken has not caused the whole population to die for want of food. The four principal bodily demands—eating, sleeping, mating and defending—can be satisfied under any circumstances, whether one is born in an impious condition or in a pious condition. The problem, however, is how to become free from the four principles of birth, death, old age and disease.

This is the real problem. It is not "What shall I eat?" The birds and beasts have no such problem. In the morning they are immediately chirping, "*Jee, jee, jee, jee.*" They know that they will have their food. No one is dying, and there is no such thing as overpopulation because everyone is provided for by God's arrangement. There are qualitative differences, but obtaining a superior quality of material enjoyment is not the end of life. The real problem is how to get free of birth, death, old age and disease. This cannot be solved by simply wasting time traveling within this universe. Even if one goes to the highest planet, this problem cannot be solved, for there is death everywhere.

The duration of life on the moon, according to Vedic information, is ten thousand years, and one day there is equal to six months here. Thus ten thousand multiplied by one hundred eighty years is the duration of life on the moon. However, it is impossible for earthmen to go to the moon and live there for very long. Otherwise the whole Vedic literature would be false. We can attempt to go there, but it is not possible to live there. This knowledge is in the *Vedas*. Therefore, we are not very eager to go to this planet or that planet. We are eager to go directly to the planet where Kṛṣṇa lives. Kṛṣṇa states in the *Bhagavad-gītā* (9.25):

yānti deva-vratā devān
pitṝn yānti pitṛ-vratāḥ
bhūtāni yānti bhūtejyā
yānti mad-yājino 'pi mām

"One can go to the moon, or one can even go to the sun or to millions and trillions of other planets, or if one is too materially attached he may remain here—but those who are My devotees will come to Me." This is our aim. Initiation into Kṛṣṇa consciousness insures that the student ultimately can go to the supreme planet, Kṛṣṇaloka. We are not sitting idly; we are also attempting to go to other planets, but we are not merely wasting time.

A sane and intelligent man does not wish to enter any of the material planets because the four conditions of material miseries exist on all of them. From the *Bhagavad-gītā* we can understand that even if we enter Brahmaloka, the highest planetary system of this universe, the four principles of misery will be present. We learn from the *Bhagavad-gītā* that the duration of one day on Brahmaloka is millions of years of our calculation. That is a fact.

Even the highest planetary system, Brahmaloka, may be reached, but scientists say that it will take forty thousand years at sputnik speed. Who is prepared to travel in space for forty thousand years? From the Vedic literature we can understand that we can enter any of the planets, provided we prepare for that purpose. If one prepares himself to enter into the higher planetary systems, which are said to be inhabited by demigods, he can go there. Similarly, one can go to a lower planetary system, or if one desires he can remain on this planet. Finally, if one desires, he can enter the planet of the Supreme Personality of Godhead. It is all a matter of preparation. However, all planetary systems within our material universe are temporary. The duration of life on certain material planets may be very long, but all living entities in the material universe are eventually subject to annihilation and have to again develop other bodies. There are different types of bodies. A human body exists one hundred years, whereas an insect body may exist for twelve hours. Thus the duration of these different bodies is relative. If one enters the planet called Vaikuṇṭhaloka, the spiritual planet, however, he then achieves eternal life, full of bliss and knowledge. A human being can attain that perfection if he tries. That is stated in the *Bhagavad-gītā* when the

Lord says, "Anyone who knows in truth about the Supreme Personality of Godhead can attain to My nature."

Many people claim, "God is great," but this is a hackneyed phrase. One must know *how* He is great, and that can be known from authorized scripture. In the *Bhagavad-gītā* God describes Himself. He says, "My appearance of taking birth just like an ordinary human being is actually transcendental." God is so kind that He comes before us as an ordinary human being, but His body is not exactly like a human body. Those rascals who do not know about Him think that Kṛṣṇa is like one of us. That is also stated in the *Bhagavad-gītā* (9.11):

avajānanti māṁ mūḍhā
mānuṣīṁ tanum āśritam
paraṁ bhāvam ajānanto
mama bhūta-maheśvaram

"Fools deride Me when I descend in the human form. They do not know My transcendental nature and My supreme dominion over all that be." We have a chance to know about Kṛṣṇa provided we read the right literature under the right direction, and if we simply know what the nature of God is, then by understanding this one fact alone we become liberated. It is not possible in our human condition to understand the Absolute Supreme Personality of Godhead completely, but with the help of the *Bhagavad-gītā,* the statements given by the Supreme Personality of Godhead, and of the spiritual master, we can know Him to the best of our capacity. If we can know Him in reality, then immediately after leaving this body we can enter into the kingdom of God. Kṛṣṇa says, *tyaktvā dehaṁ punar janma naiti mām eti so 'rjuna:* "After leaving this body, one who is in knowledge of Me does not come again to this material world, for he enters into the spiritual world and comes to Me." (Bg. 4.9)

The purpose of our Kṛṣṇa consciousness movement is to propagate this advanced scientific idea to people in general, and the process is very simple. Simply by chanting the holy names of God—Hare Kṛṣṇa, Hare Kṛṣṇa, Kṛṣṇa Kṛṣṇa, Hare Hare/ Hare Rāma, Hare Rāma, Rāma Rāma, Hare Hare—one cleanses the dirt from his heart and gains understanding that he is part and parcel of the Supreme Lord and that it is his duty to serve Him. This process is very pleasant: we chant the Hare Kṛṣṇa *mantra,* we dance rhythmically, and we eat nice *prasāda.* While enjoying this life, we are preparing to enter into the kingdom of God in our next life. This is not a fabrication—it is all factual. Although to a layman this appears to be a fabrication, Kṛṣṇa reveals Himself from within to one who is serious about God realization. Both Kṛṣṇa and the spiritual master help the sincere soul. The spiritual master is the external manifestation of God, who is situated in everyone's heart as the Supersoul. For one who is very serious about understanding the Supreme Personality of Godhead, the Supersoul immediately renders assistance by directing him to a bona fide spiritual master. In this way the spiritual candidate is helped from within and without.

According to the *Bhāgavata Purāṇa,* the Supreme Truth is realized in three stages. First there is impersonal Brahman, or the impersonal Absolute; then the Paramātmā, or localized aspect of Brahman. The neutron of the atom may be taken as the representation of the Paramātmā, who also enters into the atom. This is described in the *Brahma-saṁhitā.* But ultimately the Supreme Divine Being is realized as the supreme all-attractive person (Kṛṣṇa) with full and inconceivable potencies of opulence, strength, fame, beauty, knowledge and renunciation. These six potencies are fully exhibited by Śrī Rāma and Śrī Kṛṣṇa when They descend before human beings. Only a section of human beings—the unalloyed devotees—can recognize Kṛṣṇa on the authority of revealed scriptures, but others are bewildered by the influence of the material energy. The Absolute Truth is therefore the Absolute Person, who has no equal or competitor. The impersonal Brahman rays are the rays of His transcendental body, just as the sun's rays are emanations from the sun.

According to the *Viṣṇu Purāṇa,* the material energy is called *avidyā,* or nescience, and is exhibited in the fruitive activities of sense enjoyment. But although the living being has the tendency to be illusioned and trapped by the material energy for sense enjoyment, he belongs to the antimaterial energy, or spiritual energy. In this sense the living being is the positive energy, whereas matter is the negative energy. Matter does not develop unless in contact with the superior spiritual, or antimaterial, energy, which is directly part and parcel of the spiritual whole. The subject matter of this spiritual energy exhibited by living beings is undoubtedly very complicated for an ordinary man, who is therefore astounded by the subject. Sometimes he partially understands it through his imperfect senses, and sometimes he fails to know it altogether. It is best, therefore, to hear from the highest authority, Śrī Kṛṣṇa, or from His devotee who represents Him in the chain of disciplic succession.

This Kṛṣṇa consciousness movement is meant for the purpose of understanding God. The spiritual master is the living representative of Kṛṣṇa who helps externally, and Kṛṣṇa as the Supersoul helps internally. The living entity can take advantage of such guidance and make his life successful. We request that everyone read authoritative literature in order to understand this movement. We have published *Bhagavad-gītā As It Is; Teachings of Lord Caitanya; Śrīmad-Bhāgavatam; Kṛṣṇa, the Supreme Personality of Godhead;* and *The Nectar of Devotion.** We are also publishing our magazine *Back to*

---

*See page ii for more books by the author.

Godhead every month in many languages.

Everyone should attempt to go to Kṛṣṇa. We have published an article in our *Back to Godhead* magazine entitled "Beyond the Universe." This article describes a place beyond this universe according to knowledge which is in the *Bhagavad-gītā*. The *Bhagavad-gītā* is a very popular book, and there are many editions of it in America and also many from India. Unfortunately, however, many rascals have come to the West to preach the *Bhagavad-gītā*. They are designated as rascals because they are bluffers who do not give real information. In our *Bhagavad-gītā As It Is*, however, the spiritual nature is authoritatively described.

This cosmic manifestation is called "nature," but there is another nature, which is superior. The cosmic manifestation is inferior nature, but beyond this nature, which is manifested and unmanifested, there is another nature, which is called *sanātana*, eternal. It is easy to understand that everything manifested here is temporary. The obvious example is our body. If one is thirty years old, thirty years ago his body was not manifested, and in another fifty years it will again be unmanifested. That is a factual law of nature. It is manifested and again annihilated, just as waves in the sea rise frequently and then recede. The materialist, however, is simply concerned with this mortal life, which can be finished at any moment. Furthermore, as this body will die, so the entire universe, this gigantic material body, will be annihilated, and whether we are fortunate or unfortunate, on this planet or another planet, everything will be finished. Why then are we wasting our time trying to go to a planet where everything will be finished? We should try to go to Kṛṣṇaloka. This is spiritual science; we must try to understand it, and, after understanding it ourselves, we should preach this message to the whole world. Everyone is in darkness. Although people have no knowledge, they are very proud. But it is not advancement of knowledge to go to the moon after ten years of effort and take a rock and come back. The space travelers are very proud: "Oh, I have touched it." But what have they gained? Even if we were able to live there, it would not be for long. It will all be destroyed in the end.

Try to find that planet from which one will never return, where there is eternal life, and where one can dance with Kṛṣṇa. This is the meaning of Kṛṣṇa consciousness. Take this movement seriously, for Kṛṣṇa consciousness gives one a chance to reach Kṛṣṇa and to dance with Him eternally. From the Vedic literature we understand that this material world is a manifestation of only one fourth of the complete creation of God. The three-fourths portion of God's creation is the spiritual world. That we find in the *Bhagavad-gītā*. Kṛṣṇa says, "This material world is but a fractional part of the whole." If we look as far as we can see—up to the sky—our vision is still confined within only one universe, and there are unlimited universes clustered together within what is called the material world. But beyond those clusters of unlimited numbers of universes is the spiritual sky, which is also mentioned in the *Bhagavad-gītā,* where the Lord says that beyond the material world is another nature, which is eternal; there is no history of its beginning, and it has no end. "Eternal" refers to that which has no end and no beginning. The Vedic religion is therefore called eternal because no one can trace back when it began. The Christian religion has a history of two thousand years, and the Muhammadan religion also has a history, but if one were to trace back Vedic religion, he would not find its historical beginning. Therefore it is called eternal religion.

We may rightly say that God created this material world, and this indicates that God existed before the creation. This very word "created" suggests that before the creation of the cosmic manifestation, the Lord was existing. Therefore God is not under the creation. If God were under the creation, how could He have created it? He would instead have been one of the objects of this material creation. But God is not under the creation; He is the creator, and therefore He is eternal.

There is a spiritual sky, where there are innumerable spiritual planets and innumerable spiritual living entities, but those who are not fit to live in that spiritual world are sent to this material world. Voluntarily we have accepted this material body, but actually we are spirit souls who should not have accepted it. When and how we accepted it cannot be traced. No one can trace the history of when the conditioned soul first accepted the material body. There are 8,400,000 forms of living entities. Nine hundred thousand species of living entities are within the water; two million species of life are among the plants and vegetables. Unfortunately, this Vedic knowledge is not instructed by any university. But these are facts. Let the botanist and anthropologist research into the Vedic conclusion. Darwin's theory of the evolution of organic matter is, of course, very prominent in the institutions of learning. But the *Bhāgavata Purāṇa* and other authoritative scriptures of scientific magnitude describe how the living entities in different forms of body evolve one after another. It is not a new idea, but educators are giving stress only to Darwin's theory, although in Vedic literature we have immense information of the living conditions in this material world.

We are only a fractional portion of all the living entities in the many universes of the material world. Those who are in a material body in the material world are condemned. For example, the population in prison is condemned by the government, but their number is only a fraction of the whole population. It is not that the whole population goes to prison; some, who are disobedient, are confined in prison. Similarly, the conditioned souls within this material world are only a fraction of all the living entities in the creation of God, and because they have disobeyed God—because they did not abide by the order of Kṛṣṇa—they have been put into this material world. If one is sensible and inquisitive, he should try to understand: "Why have I been put into this conditioned life? I do not wish to suffer."

There are three kinds of suffering, one of which includes miseries pertaining to the body and mind. In Hawaii, in front of my house, a man was keeping some animals and birds for the purpose of taking them to be slaughtered. I gave this example to my students: "These animals are standing here, and if you tell them, 'Oh, my dear animals, why are you standing here? Go away! You are meant for the slaughterhouse,' they cannot go. They have no intelligence."

Suffering without knowledge, without remedy, is animal life. One who cannot understand that he is suffering and who thinks that he is very well off is in animal consciousness, not human consciousness. The human being should be cognizant of suffering the threefold miseries of this planet. One should know that he is suffering in birth, suffering in death, suffering in old age and suffering in disease, and one should be inquisitive as to how he may avoid the suffering. That is real research work.

We have suffered from the beginning of our birth. As a baby, the human being is tightly placed in the abdomen of the mother in an airtight bag for nine months. He cannot even move, there are insects biting him, and he cannot protest. After the child comes out, the suffering continues. The mother undoubtedly takes much care, but still the child cries because he is suffering. There are bugs biting or there are pains in his stomach; the child is crying, and the mother does not know how to pacify him. His suffering begins in the womb of his mother. Then, after his birth, as he grows up, there is more suffering. He does not want to go to school, but he is forced to.

He does not want to study, but the teacher gives him tasks. If we analyze our life, we will find that it is full of suffering. Why then are we coming here? The conditioned souls are not very bright. We should inquire, "Why am I suffering?" If there is a remedy, we must take advantage of it.

We are eternally connected with the Supreme Lord, but somehow or other we are now in material contamination. Therefore, we must take up a process by which to go back again to the spiritual world. That linking process is called *yoga*. The actual translation of the word *yoga* is "plus." At the present moment we are minus God, or minus the Supreme. But when we make ourselves plus—connected—then our human form of life is perfect. During our lifetime we have to practice approaching that point of perfection, and at the time of death, when we give up this material body, that perfection has to be realized. At the time of death, one must be prepared. Students, for instance, prepare for two to five years in college, and the final test of their education is the examination. If they pass the examination, they get a degree. Similarly, in the subject of life, if we prepare for the examination at the time of death and pass it, then we are transferred to the spiritual world. Everything is examined at the time of death.

There is a very common Bengali proverb that says that whatever one does for perfection will be tested at the time of his death. The *Bhagavad-gītā* describes what we should do at the point of our death, when we are giving up this present body. For the *dhyāna-yogī* (meditator) Śrī Kṛṣṇa speaks the following verses:

yad akṣaraṁ veda-vido vadanti
viśanti yad yatayo vīta-rāgāḥ
yad icchanto brahmacaryaṁ caranti
tat te padaṁ saṅgraheṇa pravakṣye

sarva-dvārāṇi saṁyamya
mano hṛdi nirudhya ca
mūrdhny ādhāyātmanaḥ prāṇam
āsthito yoga-dhāraṇām

"Persons learned in the *Vedas,* who utter *oṁkāra,* and who are great sages in the renounced order enter into Brahman. Desiring such perfection, one practices celibacy. I shall now explain to you this process by which one may attain salvation. The yogic situation is that of detachment from all sensual engagements. Closing all the doors of the senses and fixing the mind on the heart and the life air at the top of the head, one establishes himself in *yoga.*" (Bg. 8.11–12) In the *yoga* system this process is called *pratyāhāra,* which means, in technical language, "the opposite." Now the eyes are engaged in seeing worldly beauty, so one has to withdraw them from enjoying that beauty and concentrate on seeing beauty inside. That is called *pratyāhāra.* Similarly, one has to hear the *oṁkāra* sound from within.

oṁ ity ekākṣaraṁ brahma
vyāharan mām anusmaran
yaḥ prayāti tyajan dehaṁ
sa yāti paramāṁ gatim

"After being situated in this *yoga* practice and vibrating the sacred syllable *oṁ,* the supreme combination of letters, if one thinks of the Supreme Personality of Godhead and quits his body, he will certainly reach the spiritual planets." (Bg. 8.13) In this way all the senses have to be stopped in their external activities, and the mind must be concentrated on *viṣṇu-mūrti,* the form of Lord Viṣṇu. That is the perfection of *yoga.* The mind is very turbulent, so it has to be fixed upon the heart. When the mind is fixed within the heart and the life air is transferred to the top of the head, one can attain the perfection of *yoga*.

The perfect *yogī* then determines where he is to go. There are innumerable material planets, and beyond these planets there is the spiritual world. *Yogīs* have this information from Vedic scriptures. For example, before I came to the United States I read descriptions of it from books. Similarly, a description of the higher planets and the spiritual world can be found in the Vedic scriptures. The *yogī* knows everything; he can transfer himself to any planet he likes. He does not need the help of a spacecraft.

Material scientists have been trying for many years, and they will go on trying for one hundred or one thousand years more, but they will never reach any planet. Maybe by a scientific process one or two men can reach some planet, but that is not the general process. The generally accepted process for transferral to other planets is the practice of the *yoga* system or the *jñāna* system. The *bhakti* system, however, is not meant for transferral to any material planet. Those who engage in the devotional service of Kṛṣṇa, or the Supreme Lord, are not interested in any of the planets of this material world because they know that no matter to which planet one elevates himself, he will still find the four principles of material existence there nonetheless. On some planets the duration of life is much longer than on this earth, but death is there. Those who are Kṛṣṇa conscious, however, transcend this material life of birth, death, disease and old age.

Spiritual life means release from this botheration and misery. Those who are intelligent, therefore, do not try to elevate themselves to any planet of this material world. Men are trying to reach the moon, and although it is very difficult to gain entrance to that planet, if we do gain entrance the period of our lives will be enhanced. Of course, that does not apply to life in this body. If we were to enter the moon with this body, instant death would be certain.

When one enters into a planetary system, he must have a suitable body for that planet. Every planet is inhabited by living entities with bodies suitable for that planet. For instance, we can enter the water in this body, but we cannot live there. We may stay there fifteen or sixteen hours, or maybe twenty-four hours, but that's all. Aquatic animals, however, have particular bodies suitable for living their whole lives in water. Similarly, if one takes a fish out of water and puts it on the land, it will die instantly. As we understand that even on this planet there are different kinds of bodies for living in particular places, so, similarly, if we want to enter another planet, we have to prepare ourselves to get a suitable body.

If one transfers himself and his soul transmigrates to the moon by this yogic process, he gets a long duration of life. On the higher planets, six of our months equal one day. Thus the beings there live for ten thousand years. That is the description in the Vedic literature. So undoubtedly one can get a very long duration of life, but still there is death. After ten thousand or twenty thousand years, or even after millions of years (it does not matter), death comes.

Actually, we are not subject to death. That is affirmed in the beginning of the *Bhagavad-gītā* (2.20): *na hanyate hanyamāne śarīre.* We are spirit soul, and therefore we are eternal. Why then should we subject ourselves to death and birth? It is intelligent to think in this way. Those who are Kṛṣṇa conscious are very intelligent because they are not interested in getting promotion to any planet where there is death, despite a long duration of life there. Rather, they want to get a body like God's. *Īśvaraḥ paramaḥ kṛṣṇaḥ sac-cid-ānanda-vigrahaḥ.* (*Bs.* 5.1) God's body is *sac-cid-ānanda. Sat* means "eternal," and *cit* means "full of knowledge." *Ānanda* means "full of pleasure."

As stated in our pamphlet *Kṛṣṇa, the Reservoir of Pleasure*, if we transfer ourselves to the spiritual world, to Kṛṣṇa's planet or to any other spiritual planet, then we will get a body similar to God's: *sac-cid-ānanda*—eternal, full of knowledge and full of bliss. So those who try to be Kṛṣṇa conscious have a different aim of life than those who are trying to promote themselves to the better planets in this material world. Lord Kṛṣṇa says, *yaḥ prayāti tyajan dehaṁ sa yāti paramāṁ gatim:* "The perfection of *yoga* is to transfer oneself to the spiritual world." (Bg. 8.13)

The spirit soul is a minute particle within the body. We cannot see it. One practices the *yoga* system to raise the soul to the topmost part of the head. This practice goes on while one is living, and the perfection is reached when one can place himself on the top of the head and then break through. Then he can transfer himself to whatever higher planets he likes. That is the perfection of the *yogī.*

If the *yogī* is inquisitive to see the moon, he can say, "Ah, let me see what the moon is like. Then I shall transfer myself to higher planets," just like travelers who go to Europe, California, Canada, or other countries on earth. One can transfer oneself to many planets by this *yoga* system, but anywhere he goes he will find visa systems and customs systems. To go to other planets, one must be qualified.

Kṛṣṇa conscious persons are not interested in any temporary planet, even if it offers a long duration of life. If the *yogī*, at the time of death, can pronounce *oṁ*, the concise form of transcendental vibration, and at the same time remember Kṛṣṇa, Viṣṇu (*mām anusmaran*), he will attain perfection. The purpose of the entire *yoga* system is to concentrate the mind on Viṣṇu. Impersonalists imagine that they see the form of Viṣṇu, or the Lord, but those who are personalists do not imagine this—they actually see the form of the Supreme Lord.

Either way, if one concentrates his mind through imagination or if one actually sees, one has to concentrate his mind on the Viṣṇu form. *Mām* means "unto the Supreme Lord, Viṣṇu." Anyone who leaves this body and concentrates his mind on Viṣṇu enters into the spiritual kingdom after quitting his body. Those who are actually *yogīs* do not desire to enter any other planet because they know that life is temporary on the temporary planets, and thus they are not interested. That is intelligence.

Those who are satisfied with temporary happiness, temporary life and temporary facilities are not intelligent according to the *Bhagavad-gītā* (7.23). *Antavat tu phalaṁ teṣāṁ tad bhavaty alpa-medhasām:* "One whose brain substance is very meager is interested in temporary things." That is the version of *Śrīmad Bhagavad-gītā.* I am eternal, so why should I be interested in nonpermanent things? Who wants nonpermanent existence? No one wants it. If we are living in an apartment and the landlord asks us to vacate, we are sorry, but we are not sorry if we move to a better apartment. This then is our inclination. We do not wish to die, because we are eternal.

The material atmosphere is robbing us of our eternality. The *Śrīmad-Bhāgavatam* says, "Our duration of life is being diminished by the sun, beginning from its rising until the time it sets." Daily we are losing the duration of our lives. If the sun rises at 5:30 in the morning, at 5:30 in the evening twelve hours have been taken away from the duration of our lives. We will never get this time back. If we ask any scientist, "I will give you twelve million dollars—please give me back these twelve hours," he will reply, "No, it is not possible." The scientist cannot do it. Therefore the *Bhāgavatam* says that from sunrise to sunset the duration of our lives is being diminished.

Time is called *kāla*—past, present and future. What is now present, tomorrow will be past, and what is now future, tomorrow will be present. But this past, present and future are the past, present and future of the body. We do not belong to the category of the past, present and future. We belong to the category of eternity. Therefore one should be concerned with how to attain or how to be elevated to the platform of eternity. The developed consciousness of the human being should be utilized not in the animal propensities of eating, sleeping, mating and defending but in searching out the valuable path which will help him get that life of eternity. It is said that the sun is taking away our duration of life—every minute, every hour, every day—but if we engage ourselves in the topics of Uttamaśloka, the topics of the Lord, that time cannot be taken away. The time one devotes in a Kṛṣṇa consciousness temple cannot be taken away. It is an asset—a plus, not a minus. The duration of life, so far as the body is concerned, may be taken; however one tries to keep it intact, no one can do it. But the spiritual education we receive in Kṛṣṇa consciousness cannot be taken away by the sun. It becomes a solid asset.

Chanting Hare Kṛṣṇa, Hare Kṛṣṇa, Kṛṣṇa Kṛṣṇa, Hare Hare/ Hare Rāma, Hare Rāma, Rāma Rāma, Hare Hare is a very easy thing to do. Time spent chanting cannot be taken away like time pertaining to the body. Fifty years ago I was a young man, but that time has been taken and cannot be returned. The spiritual knowledge I received from my spiritual master, however, cannot be taken away, but will go with me. Even after I quit this body, it will go with me; and if it is perfect in this life, then it will take me to the eternal abode.

Both the material and spiritual worlds belong to Kṛṣṇa. We are not proprietors of anything. It is all the property of the Supreme Lord, just as everything in the state belongs to the government, either in the prison house or outside the prison house. Conditioned life is just like life in a prison house in this material world. A prisoner cannot freely change from one cell to another. In free life one can go from one home to another home, but in prison life one cannot do that but must stay in his cell. All these planets are like cells. We are trying to go to the moon, but it is not practical by mechanical means. Whether we are American, Indian, Chinese or Russian, we have been given this planet to live on. We cannot leave—although there are millions and billions of planets and although we have so many flying machines—because we are conditioned by the laws of nature, God's laws. A man who is put into a certain cell cannot change at will without superior authority. Kṛṣṇa says in the *Bhagavad-gītā* that one should not try to change from one cell to another. That will not make anyone happy. If a prisoner thinks, "I am in this cell—let me request the warden to change my cell, and I will be happy," that is a mistaken idea. One cannot be happy so long as he is within the prison walls. We are trying to be happy by changing cells—from capitalism to communism or communism to capitalism. The aim should be to become free from this "ism" and that "ism." One has to change completely from the "ism" called materialism; then he can become happy. That is the program of Kṛṣṇa consciousness.

We are taking advice from the Supreme Person. He says, "My dear Arjuna, you may be elevated to the highest planetary system, which is called Brahmaloka and is desirable because life there is very long." We cannot calculate even a half-day there. It is beyond our mathematical calculations. But even in Brahmaloka there is death. Therefore Kṛṣṇa says, "Do not waste your time trying to elevate yourself or transfer yourself from this planet to that planet."

The people I have seen in America are very restless. They go from one apartment to another apartment or from one country to another country. That restlessness is there because we are searching after our real home. To go from this place to that place will not give eternal life. Eternal life is with Kṛṣṇa. Therefore Kṛṣṇa says, "Everything belongs to Me, and I have the superexcellent abode, which is called Goloka Vṛndāvana." If one wants to go there, he must simply become Kṛṣṇa conscious and try to understand how Kṛṣṇa appears and disappears, what His constitutional position is, what our constitutional position is, what our relationship with Him is, and how to live in that relationship. Simply try to understand these ideas scientifically. Everything in Kṛṣṇa consciousness is scientific. It is not bogus, whimsical, sentimental, fanatical or imaginary. It is truth, fact, reality. One must understand Kṛṣṇa in truth.

We have to give up this body, willingly or unwillingly. The day will come when we will have to submit to the laws of nature and give up this body. Even President Kennedy in his procession had to submit to nature's law and change his body for another body. He could not say, "Oh, I am the President; I am Mr. Kennedy. I cannot do that." He was forced to do it. That is the way nature works.

The purpose of our developed human consciousness is to understand how nature works. Aside from human consciousness, there is consciousness in dogs, cats, worms, trees, birds, beasts and all other species. But we are not meant to live in that consciousness. The *Śrīmad-Bhāgavatam* says that after many, many births we have attained the human form of body. Now we should not misuse it. Please utilize this human life to develop Kṛṣṇa consciousness and be happy.

Glossary

Aparā prakṛti—the inferior, material energy of the Lord.

Aṣṭāṅga-yoga—the eightfold system of mystic yoga, propounded by Patañjali, meant for realizing the presence of Paramātmā, the Lord in the heart.

Avidyā—nescience, ignorance.

Bhagavān—the Supreme Personality of Godhead, who possesses in full the opulences of wealth, beauty, strength, knowledge, fame and renunciation.

Bhakti-yoga—loving devotional service unto Kṛṣṇa; the activity of the superior, spiritual energy.

Bhāva—the stage just prior to love of Godhead; emotion on the spiritual plane, which is transcendental to mental and intellectual functions.

Brahmā— the first created living being and secondary creator of the material universe; the presiding deity of this universe.

Brahman—the effulgence emanating from the transcendental body of Lord Kṛṣṇa; the impersonal Absolute.

Brahma-jyoti—the spiritual bodily effulgence of Kṛṣṇa.

Brahma-pāda—occupation of the post of Brahmā.

Brahma-randhra—the hole on the topmost part of the skull. By lifting the vital force to the *brahma-randhra,* a *yogī* can leave the gross and subtle bodies and reach the transcendental Vaikuṇṭha planets.

Dhūma—the time of the smoke. A *yogī* who dies during this time reaches the moon but again comes back.

Guṇas—the three modes of nature.

Hare—a form of address to the internal energy of the Lord.

Jīva—the living spirit, or vital force.

Kāla—eternal time.

Parā prakṛti—the Lord's superior energy, which creates the antimaterial world.

Paravyoma—the variegated spiritual planetary system that comprises three fourths of the Supreme Lord's energy. Also called Vaikuṇṭhaloka.

Pitās—forefathers.

Pratyāhāra—the process of closing all the doors of the senses and fixing the mind on the heart and the life air at the top of the head, thus establishing oneself in *yoga*.

Rajas—the material mode of passion.

Sanātana—eternal, that which has neither beginning nor end.

Sanātana-dhāma—the eternal nature, the antimaterial sky beyond the material universe.

Sanātana-dharma—the eternal nature of the living being, to render service.

Sāṅkhyaites—speculators who scrutinize the material principles with minute analysis and synthesis.

Sattva—the material mode of goodness.

Satyaloka—the topmost planet of the material world. Also called Brahmaloka.

Siddhaloka—planets of materially perfect beings who have full capacities to control gravity, space, time, etc.

Soma-rasa—a celestial beverage drunk on the moon.

Tamas—the material mode of ignorance.

Sanskrit Pronunciation Guide

The system of transliteration used in this book conforms to a system that scholars have accepted to indicate the pronunciation of each sound in the Sanskrit language.

The short vowel **a** is pronounced like the **u** in but, long **ā** like the **a** in far. Short **i** is pronounced as in pin, long **ī** as in pique, short **u** as in pull, and long **ū** as in rule. The vowel **ṛ** is pronounced like the **ri** in **rim**, **e** like the **ey** in they, **o** like the **o** in go, **ai** like the **ai** in **aisle**, and **au** like the **ow** in how. The *anusvāra* (**ṁ**) is pronounced like the **n** in the French word *bon*, and *visarga* (**ḥ**) is pronounced as a final **h** sound. At the end of a couplet, **aḥ** is pronounced **aha**, and **iḥ** is pronounced **ihi**.

The guttural consonants—**k, kh, g, gh,** and **ṅ**—are pronounced from the throat in much the same manner as in English. **K** is pronounced as in **kite**, **kh** as in Eck**h**art, **g** as in **g**ive, **gh** as in di**g h**ard, and **ṅ** as in si**ng**.

The palatal consonants—**c, ch, j, jh,** and **ñ**—are pronounced with the tongue touching the firm ridge behind the teeth. **C** is pronounced as in **ch**air, **ch** as in staun**ch-h**eart, **j** as in **j**oy, **jh** as in he**dgeh**og, and **ñ** as in ca**ny**on.

The cerebral consonants—**ṭ, ṭh, ḍ, ḍh,** and **ṇ**—are pronounced with the tip of the tongue turned up and drawn back against the dome of the palate. **Ṭ** is pronounced as in **t**ub, **ṭh** as in ligh**t-h**eart, **ḍ** as in **d**ove, **ḍh** as in re**d-h**ot, and **ṇ** as in **n**ut. The dental consonants—**t, th, d, dh,** and **n**—are pronounced in the same manner as the cerebrals, but with the forepart of the tongue against the teeth.

The labial consonants—**p, ph, b, bh,** and **m**—are pronounced with the lips. **P** is pronounced as in **p**ine, **ph** as in u**ph**ill, **b** as in **b**ird, **bh** as in ru**b-h**ard, and **m** as in **m**other.

The semivowels—**y, r, l,** and **v**—are pronounced as in **y**es, **r**un, **l**ight, and **v**ine respectively. The sibilants—**ś, ṣ,** and **s**—are pronounced, respectively, as in the German word *sprechen* and the English words **sh**ine and **s**un. The letter **h** is pronounced as in **h**ome.

The Author

His Divine Grace A. C. Bhaktivedanta Swami Prabhupāda appeared in this world in 1896 in Calcutta, India. He first met his spiritual master, Śrīla Bhaktisiddhānta Sarasvatī Gosvāmī, in Calcutta in 1922. Bhaktisiddhānta Sarasvatī, a prominent religious scholar and the founder of sixty-four Gauḍīya Maṭhas (Vedic institutes), liked this educated young man and convinced him to dedicate his life to teaching Vedic knowledge. Śrīla Prabhupāda became his student and, in 1933, his formally initiated disciple.

At their first meeting, in 1922, Śrīla Bhaktisiddhānta Sarasvatī requested Śrīla Prabhupāda to broadcast Vedic knowledge in English. In the years that followed, Śrīla Prabhupāda wrote a commentary on the *Bhagavad-gītā*, assisted the Gauḍīya Maṭha in its work, and, in 1944, started *Back to Godhead*, an English fortnightly magazine. Single-handedly, Śrīla Prabhupāda edited it, typed the manuscripts, checked the galley proofs, and even distributed the individual copies. The magazine is now being continued by his followers.

In 1950 Śrīla Prabhupāda retired from married life, adopting the *vānaprastha* (retired) order to devote more time to his studies and writing. He traveled to the holy city of Vṛndāvana, where he lived in humble circumstances in the historic temple of Rādhā-Dāmodara. There he engaged for several years in deep study and writing. He accepted the renounced order of life (*sannyāsa*) in 1959. At Rādhā-Dāmodara, Śrīla Prabhupāda began work on his life's masterpiece: a multivolume commentated translation of the eighteen-thousand-verse *Śrīmad-Bhāgavatam* (*Bhāgavata Purāṇa*). He also wrote *Easy Journey to Other Planets*.

After publishing three volumes of the *Bhāgavatam*, Śrīla Prabhupāda came to the United States, in September 1965, to fulfill the mission of his spiritual master. Subsequently, His Divine Grace wrote more than fifty volumes of authoritative commentated translations and summary studies of the philosophical and religious classics of India.

When he first arrived by freighter in New York City, Śrīla Prabhupāda was practically penniless. Only after almost a year of great difficulty did he establish the International Society for Krishna Consciousness, in July of 1966. Before he passed away on November 14, 1977, he had guided the Society and seen it grow to a worldwide confederation of more than one hundred *āśramas*, schools, temples, institutes, and farm communities.

In 1972 His Divine Grace introduced the Vedic system of primary and secondary education in the West by founding the *gurukula* school in Dallas, Texas. Since then his disciples have established similar schools throughout the United States and the rest of the world.

Śrīla Prabhupāda also inspired the construction of several large international cultural centers in India. The center at Śrīdhāma Māyāpur is the site for a planned spiritual city, with the magnificent Temple of the Vedic Planetarium at its heart. In Vṛndāvana are the Kṛṣṇa-Balarāma Temple and International Guesthouse, *gurukula* school, and Śrīla Prabhupāda Memorial and Museum. There are also major cultural and educational centers in Mumbai, New Delhi, Baroda, Siliguri, Ujjain, and Ahmedabad. Other centers are either underway or planned in a dozen important locations on the Indian subcontinent.

Śrīla Prabhupāda's most significant contribution, however, is his books. Highly respected by scholars for their authority, depth, and clarity, they are used as textbooks in numerous college courses. His writings have been translated into over fifty languages. The Bhaktivedanta Book Trust, established in 1972 to publish the works of His Divine Grace, has thus become the world's largest publisher of books in the field of Indian religion and philosophy.

In just twelve years, despite his advanced age, Śrīla Prabhupāda circled the globe fourteen times on lecture tours that took him to six continents. In spite of such a vigorous schedule, Śrīla Prabhupāda continued to write prolifically. His writings constitute a veritable library of Vedic philosophy, religion, literature, and culture.

The International Society for Krishna Consciousness
Founder-*ācārya:* His Divine Grace A.C. Bhaktivedanta Swami Prabhupāda

CENTERS AROUND THE WORLD

(Partial List)

NORTH AMERICA

CANADA

Brampton-Mississauga, Ontario — Unit 20, 1030 Kamato Dr., L4W 4B6/ Tel. (416) 840-6587 or (905) 826-1290/ iskconbrampton@gmail.com

Calgary, Alberta — 313 Fourth St. N.E., T2E 3S3/ Tel. (403) 265-3302/ vamanstones@shaw.ca

Edmonton, Alberta — 9353 35th Ave. NW, T6E 5R5/ Tel. (780) 439-9999/ edmonton@harekrishnatemple.com

Montreal, Quebec — 1626 Pie IX Boulevard, H1V 2C5/ Tel. & fax: (514) 521-1301/ iskconmontreal@gmail.com

♦ **Ottawa, Ontario** — 212 Somerset St. E., K1N 6V4/ Tel. (613) 565-6544/ radha_damodara@yahoo.com

Regina, Saskatchewan — 1279 Retallack St., S4T 2H8/ Tel. (306) 525-0002 0r -6461/ jagadishadas@yahoo.com

Scarborough, Ontario — 3500 McNicoll Ave., Unt #3, M1V 4C7/ Tel. (416) 300 7101/ iskconscarborough@hotmail.com

♦ **Toronto, Ontario** — 243 Avenue Rd., M5R 2J6/ Tel. (416) 922-5415/ info@torontokrishna.com

♦ **Vancouver, B.C.** — 5462 S.E. Marine Dr., Burnaby V5J 3G8/ Tel. (604) 433-9728/ ISKCONVancouver@gmail.com/ Govinda's Bookstore & Cafe: (604) 433-7100 or (888) 433-8722

RURAL COMMUNITY

Ashcroft, B.C. — Saranagati Dhama, Venables Valley (mail: P.O. Box 99, V0K 1A0)/ Tel. (250) 457-7438/ info@saranagativillage.com

U.S.A.

Atlanta, Georgia — 1287 South Ponce de Leon Ave., N.E., 30306/ Tel. & fax: (404) 377-8680/ admin@atlantaharekrishnas.com

Austin, Texas — 10700 Jonwood Way, 78753/ Tel. (512) 835-2121/ sda@backtohome.com

Baltimore, Maryland — 200 Bloomsbury Ave., Catonsville, 21228/ Tel. (410) 744-1624/ contact@iskconbaltimore.org

Berkeley, California — 2334 Stuart Street, 94705/ Tel. (510) 540-9215/ info@iskconberkeley.net

Boise, Idaho — 1615 Martha St., 83706/ Tel. (208) 344-4274/ boise_temple@yahoo.com

Boston, Massachusetts — 72 Commonwealth Ave., 02116/ Tel. (617) 247-8611/ info@iskconboston.org

♦ **Chicago, Illinois** — 1716 W. Lunt Ave., 60626/ Tel. (773) 973-0900/ chicagoiskcon@yahoo.com

Columbus, Ohio — 379 W. Eighth Ave., 43201/ Tel. (614) 421-1661/ premvilasdas.rns@gmail.com

♦ **Dallas, Texas** — 5430 Gurley Ave., 75223/ Tel. (214) 827-6330/ info@radhakalachandji.com

♦ **Denver, Colorado** — 1400 Cherry St., 80220/ Tel. (303) 333-5461/ info@krishnadenver.com

Detroit, Michigan — 383 Lenox Ave., 48215/ Tel. (313) 824-6000/ gaurangi108@hotmail.com

Gainesville, Florida — 214 N.W. 14th St., 32603/ Tel. (352) 336-4183/ kalakantha.acbsp@pamho.net

Hartford, Connecticut — 1683 Main St., E. Hartford, 06108/ Tel. & fax: (860) 289-7252/ pyari108@gmail.com

Hillsboro, Oregon — 612 N 1st Ave. 97124 / Tel. : (503) 675-5000/ info@iskconportland.com

♦ **Honolulu, Hawaii** — 51 Coelho Way, 96817/ Tel. (808) 595-4913/ hawaii.iskcon@gmail.com

Houston, Texas — 1320 W. 34th St., 77018/ Tel. (713) 686-4482/ management@iskconhouston.org

Kansas City, Missouri — Rupanuga Vedic College, 5201 Paseo Blvd., 64110/ Tel. (816) 924-5640/ rvc@rvc.edu

Laguna Beach, California — 285 Legion St., 92651/ Tel. (949) 494-7029/ info@lagunatemple.com

Las Vegas, Nevada — 7181 Dean Martin Dr., 89118/ Tel. (702) 434-8332/ info@govindascenter.com

♦ **Los Angeles, California** — 3764 Watseka Ave., 90034/ Tel. (310) 836-2676/ membership@harekrishnala.com

♦ **Miami, Florida** — 3220 Virginia St., 33133 (mail: 3109 Grand Ave., #491, Coconut Grove, FL 33133/ Tel. (305) 461-1348/ devotionalservice@iskcon-miami.org

Mountain View, California — 1965 Latham St., 94040/ Tel. (650) 336-7993 / isvtemple108@gmail.com

New Orleans, Louisiana — 2936 Esplanade Ave., 70119/ Tel. (504) 304-0032 (office) or (504) 638-1944 (temple)/ gopal211@aol.com

New York, New York — 305 Schermerhorn St., Brooklyn, 11217/ Tel. (718) 855-6714/ ramabhadra@aol.com

New York, New York — The Bhakti Center, 25 First Ave., 10003/ Tel. (212) 253-6182

Orlando, Florida — 2651 Rouse Rd., 32817/ Tel. (407) 257-3865/ info@iskconorlando.com

Philadelphia, Pennsylvania — 41 West Allens Lane, 19119/ Tel. (215) 247-4600/ info@iskconphiladelphia.com

♦ Temples with restaurants or dining

Philadelphia, Pennsylvania — 1408 South St., 19146/ Tel. (215) 985-9303/ govindasvegetarian@gmail.com
Phoenix, Arizona — 100 S. Weber Dr., Chandler, 85226/ Tel. (480) 705-4900/ premadhatridd@gmail.com
◆ **St. Louis, Missouri** — 3926 Lindell Blvd., 63108/ Tel. (314) 535-8085 or 255-2207/ root@iskconstlouis.org
Salt Lake City, Utah — 965 E. 3370 South, 84106/ Tel. (801) 487-4005/ utahkrishnas@gmail.com
San Antonio, Texas — 6538 Thunderbird Dr., 78240 / Tel. (210) 420-1182/ ashram5monks@gmail.com
San Diego, California — 1030 Grand Ave., Pacific Beach, 92109/ Tel. (858) 429-9375/ krishna.sandiego@gmail.com
Seattle, Washington — 1420 228th Ave. S.E., Sammamish, 98075/ Tel. (425) 246-8436/ info@vedicculturalcenter.org
◆ **Spanish Fork, Utah** — Krishna Temple Project & KHQN Radio, 8628 S. State Road, 84660/ Tel. (801) 798-3559/ utahkrishnas@gmail.com
Tallahassee, Florida — 1323 Nylic St., 32304/ Tel. & fax: (850) 224-3803/ tallahassee.iskcon@gmail.com
Towaco, New Jersey — 100 Jacksonville Rd. (mail: P.O. Box 109), 07082/ Tel. & fax: (973) 299-0970/ madhupati.jas@pamho.net
◆ **Tucson, Arizona** — 711 E. Blacklidge Dr., 85719/ Tel. (520) 792-0630/ sandaminidd@cs.com
Washington, D.C. — 10310 Oaklyn Dr., Potomac, Maryland 20854/ Tel. (301) 299-2100/ info@iskconofdc.org

RURAL COMMUNITIES

Alachua, Florida (New Raman Reti) — 17306 N.W. 112th Blvd., 32615 (mail: P.O. Box 819, 32616)/ Tel. (386) 462-2017/ alachuatemple@gmail.com
Carriere, Mississippi (New Talavan) — 31492 Anner Road, 39426/ Tel. (601) 749-9460 or 799-1354/ talavan@hughes.net
Gurabo, Puerto Rico (New Govardhana Hill) — Carr. 181, Km. 16.3, Bo. Santa Rita, Gurabo (mail: HC-01, Box 8440, Gurabo, PR 00778)/ Tel. & fax: (787) 767-3530 or 737-1722/ manoratha@gmail.com
Hillsborough, North Carolina (New Goloka) — 1032 Dimmocks Mill Rd., 27278/ Tel. (919) 732-6492/ bkgoswami@earthlink.net
◆ **Moundsville, West Virginia (New Vrindaban)** — 3759 McCrearys Ridge Rd., 26041/ Tel. (304) 843-1600 (Guesthouse extension: 111)/ mail@newvrindaban.com
Mulberry, Tennessee (Murari-sevaka) — 532 Murari Lane, 37359 Tel. (931) 759-6888/ murari_sevaka@yahoo.com
Port Royal, Pennsylvania (Gita Nagari) — 534 Gita Nagari Rd., 17082/ Tel. (717) 527-4101/ dhruva.bts@pamho.net
Sandy Ridge, North Carolina (Prabhupada Village) — 1283 Prabhupada Rd., 27046/ Tel. (336) 593-2322/ prabhupadavillage@gmail.com

ADDITIONAL RESTAURANTS

Hato Rey, Puerto Rico — Tamal Krishna's Veggie Garden, 131 Eleanor Roosevelt, 00918/ Tel. (787) 754-6959/ tkveggiegarden@aol.com

UNITED KINGDOM AND IRELAND

Belfast, Northern Ireland — Brooklands, 140 Upper Dunmurray Lane, BT17 OHE/ Tel. +44 (028) 9062 0530/ hk.temple108@gmail.com
Birmingham, England — 84 Stanmore Rd., Edgbaston B16 9TB/ Tel. +44 (121) 420 4999/ iskconbirmingham@gmail.com
Cardiff, Wales — The Soul Centre, 116 Cow-bridge Rd., Canton/ Tel. +44 (29) 2039 0391/ the.soul.centre@pamho.net
Coventry, England — Kingfield Rd., Coventry (mail: 19 Gloucester St., Coventry CV1 3BZ)/ Tel. +44 (24) 7655 2822 or 5420/ haridas.kds@pamho.net
Dublin, Ireland — 83 Middle Abbey St., Dublin 1/ Tel. +353 (1) 661 5095/ dublin@krishna.ie; Govinda's: info@govindas.ie
Leicester, England — 21 Thoresby St., North Evington, LE5 4GU/ Tel. +44 (116) 276 2587/ pradyumna.jas@pamho.net
Lesmahagow, Scotland — Karuna Bhavan, Bankhouse Rd., Lesmahagow, Lanarkshire, ML11 0ES/ Tel. +44 (1555) 894790/ karunabhavan@aol.com
◆ **London, England (city)** — 10 Soho St., W1D 3DL/ Tel. +44 (20) 7437-3662; residential /pujaris, 7439-3606; shop, 7287-0269; Govinda's Restaurant, 7437-4928/ london@pamho.net
◆ **London, England (country)** — Bhaktivedanta Manor, Dharam Marg, Hilfield Lane, Watford, Herts, WD25 8EZ/ Tel. +44 (1923) 851000/ info@krishnatemple.com; (for accommodations:) bmguesthouse@krishna.com
London, England (south) — 42 Enmore Road, South Norwood, SE25 5NG/ Tel. +44 7988857530/ krishnaprema89@hotmail.com
London, England (Kings Cross) — 102 Caledonian Rd., Kings Cross, Islington, N1 9DN/ Tel. +44 (20) 7168 5732/ foodforalluk@aol.com
Manchester, England — 20 Mayfield Rd., Whalley Range, M16 8FT/ Tel. +44 (161) 226-4416/ contact@iskconmanchester.com
Newcastle-upon-Tyne, England — 304 Westgate Rd., NE4 6AR/ Tel. +44 (191) 272 1911
◆ **Swansea, Wales** — 8 Craddock St., SA1 3EN/ Tel. +44

(1792) 468 469/ info@iskconwales.org.uk; restaurant: info@govindas.org.uk

RURAL COMMUNITIES

London, England – (contact Bhaktivedanta Manor)

Upper Lough Erne, Northern Ireland – Govindadwipa Dhama, Inisrath Island, Derrylin, Co. Fermanagh, BT92 9GN/ Tel. +44 (28) 6772 1512/ iskconbirmingham@gmail.com

ADDITIONAL RESTAURANTS

Dublin, Ireland – Govinda's, 4 Aungier St., Dublin 2/ Tel. +353 (1) 475 0309/ info@govindas.ie

AUSTRALASIA

AUSTRALIA

Adelaide – 25 Le Hunte St. (mail: P.O. Box 114, Kilburn, SA 5084)/ Tel. & fax: +61 (8) 8359-5120/ iskconsa@tpg.com.au

Brisbane – 32 Jennifer St., Seventten Mile Rocks, QLD 4073 (mail: PO Box 525, Sherwood, QLD 4075)/ Tel. +61 (7) 3376 2388/ info@iskcon.org.au

Canberra – 44 Limestone Ave., Ainslie, ACT 2602 (mail: P.O. Box 1411, Canberra, ACT 2601)/ Tel. & fax: +61 (2) 6262-6208/ iskcon@harekrishnacanberra.com

Melbourne – 197 Danks St. (mail: P.O. Box 125), Albert Park , VIC 3206/ Tel. +61 (3) 9699-5122/ melbourne@pamho.net

Perth – 155–159 Canning Rd., Kalamunda (mail: P.O. Box 201 Kalamunda 6076)/ Tel. +61 (8) 6293-1519/ perth@pamho.net

Sydney – 180 Falcon St., North Sydney, NSW 2060 (mail: P.O. Box 459, Cammeray, NSW 2062)/ Tel. +61 (2) 9959-4558/ admin@iskcon.com.au

Sydney – Govinda's Yoga and Meditation Centre, 112 Darlinghurst Rd., Darlinghurst NSW 2010 (mail: P.O. Box 174, Kings Cross 1340)/ Tel. +61 (2) 9380-5162/ sita@govindas.com.au

RURAL COMMUNITIES

Bambra, VIC (New Nandagram) – 50 Seaches Outlet, off 1265 Winchelsea Deans Marsh Rd., Bambra VIC 3241/ Tel. +61 (3) 5288-7383

Cessnock, NSW (New Gokula) – Lewis Lane (off Mount View Rd., Millfield, near Cessnock (mail: P.O. Box 399, Cessnock, NSW 2325)/ Tel. +61 (2) 4998-1800/ iskconfarm@mac.com

Murwillumbah, NSW (New Govardhana) – Tyalgum Rd., Eungella (mail: P.O. Box 687), NSW 2484/ Tel. +61 (2) 6672-6579/ ajita@in.com.au

RESTAURANTS

Brisbane – Govinda's, 99 Elizabeth St., 1st floor, QLD 4000/ Tel. +61 (7) 3210-0255

Brisbane – Krishna's Cafe, 1st Floor, 82 Vulture St., West End, QLD 4000/ brisbane@pamho.net

Burleigh Heads – Govindas, 20 James St., Burleigh Heads, QLD 4220/ Tel. +61 (7) 5607-0782/ ajita@in.com.au

Maroochydore – Govinda's Vegetarian Cafe, 2/7 First Avenue, QLD 4558/ Tel. +61 (7) 5451-0299

Melbourne – Crossways, 1st Floor, 123 Swanston St., VIC 3000/ Tel. +61 (3) 9650-2939

Melbourne – Gopal's, 139 Swanston St., VIC 3000/ Tel. +61 (3) 9650-1578

Newcastle – 110 King Street, NSW 2300/ Tel. +61 (02) 4929-6900/ info@govindascafe.com.au

Perth – Govinda's Restaurant, 194 William St., Northbridge, W.A. 6003/ Tel. +61 (8) 9227-1648/ perth@pamho.net

Perth – Hare Krishna Food for Life, NSW 2300/ Tel. +61 (02) 4929-6900/ info@govindascafe.com.au

NEW ZEALAND AND FIJI

Christchurch, NZ – 83 Bealey Ave. (mail: P.O. Box 25-190)/ Tel. +64 (3) 366-5174/ iskconchch@clear.net.nz

Hamilton, NZ – 188 Maui St., RD 8, Te Rapa/ Tel. +64 (7) 850-5108/ rmaster@wave.co.nz

Labasa, Fiji – Delailabasa (mail: P.O. Box 133)/ Tel. +679 812912

Lautoka, Fiji – 5 Tavewa Ave. (mail: P.O. Box 125)/ Tel. +679 6664112/ regprakash@excite.com

Nausori, Fiji – Hare Krishna Cultural Centre, 2nd Floor, Shop & Save Building, 11 Gulam Nadi St., Nausori Town (mail: P.O. Box 2183, Govt. Bldgs., Suva)/ Tel. +679 9969748 or 3475097/ vdas@frca.org.fj

Rakiraki, Fiji – Rewasa (mail: P.O. Box 204)/ Tel. +679 694243

Sigatoka, Fiji – Sri Sri Radha Damodar Temple, Off Mission St., Sigatoka Town/ Tel. +679 9373703/ drgsmarna@connect.com.fj

Suva, Fiji – 166 Brewster St. (mail: P.O. Box 4299, Samabula)/ Tel. +679 3318441/ iskconsuva@connect.com.fj

Wellington, NZ – 105 Newlands Rd., Newlands/ **Tel.** +64 (4) 478-4108/ info@iskconwellington.org.nz

Wellington, NZ – Gaura Yoga Centre, 1st Floor, 175 Vivian St. (mail: P.O. Box 6271, Marion Square)/ Tel. +64 (4) 801-5500/ yoga@gaurayoga.co.nz

RURAL COMMUNITY

Auckland, NZ (New Varshan) – Hwy. 28, Riverhead, next to Huapai Golf Course (mail: R.D. 2, Kumeu)/ Tel. +64 (9) 412-8075/

RESTAURANT

Wellington, NZ – Higher Taste Hare Krishna Restaurant,

Old Bank Arcade, Ground Flr., Corner Customhouse, Quay & Hunter St., Wellington/ Tel. +64 (4) 472-2233

INDIA (partial list)*

Ahmedabad, Gujarat — Satellite Rd., Gandhinagar Highway Crossing, 380 054/ Tel. (079) 686-1945, -1645, or -2350/ jasomatinandan.acbsp@pamho.net

Allahabad, UP — Hare Krishna Dham, 161 Kashi Raj Nagar, Baluaghat 211 003/ Tel. (0532) 415294

Bangalore, Karnataka — Hare Krishna Hill, Chord Rd., 560 010/ Tel. (080) 23471956 or 23578346/ Fax: (080) 23578625/ manjunath36@iskconbangalore.org

Bangalore, Karnataka — ISKCON Sri Jagannath Mandir, No.5 Sripuram, 1st cross, Sheshadripuram, Bangalore 560 020/ Tel. (080) 3536867 or 2262024 or 3530102

Baroda, Gujarat — Hare Krishna Land, Gotri Rd., 390 021/ Tel. (0265) 2310630 or 2331012/ iskcon.baroda@pamho.net

• **Bhubaneswar, Orissa** — N.H. No. 5, IRC Village, 751 015/ Tel. (0674) 2553517, 2553475, or 2554283

Chennai (Madras), TN — Hare Krishna Land, Bhaktivedanta Swami Road, Off ECR Road, Injam- bakkam, Chennai 600 041/ Tel. (044) 5019303 or 5019147/ iskconchennai@eth.net

Dwarka, Gujarat — Bharatiya Bhavan, Devi Bhavan Rd., 361 335/ Tel. (02892) 34606/ Fax: (02892) 34319

Haridwar, Uttaranchal — Prabhupada Ashram, G. House, Nai Basti, Mahadev Nagar, Bhimgoda/ Tel. (01334) 260818

Hyderabad, AP — Hare Krishna Land, Nampally Station Rd., 500 001/ Tel. (040) 24744969 or 24607089/ iskcon.hyderabad@pamho.net

Kolkata (Calcutta), WB — 3C Albert Rd., 700 017 (behind Minto Park, opp. Birla High School)/ Tel. (033) 3028-9258 or -9280/ iskcon.calcutta@pamho.net

• **Mayapur, WB** — ISKCON, Shree Mayapur Chandrodaya Mandir, Shree Mayapur Dham, Dist. Nadia, 741 313/ Tel. (03472) 245239, 245240, or 245233/ Fax: (03472) 245238/ mayapur.chandrodaya@pamho.net

• **Mumbai (Bombay), Maharashtra** — Hare Krishna Land, Juhu 400 049/ Tel. (022) 26206860/ Fax: (022) 26205214/ info@iskconmumbai.com; guest.house.bombay@pamho.net

• **Mumbai, Maharashtra** — 7 K. M. Munshi Marg, Chowpatty 400 007 / Tel. (022) 23665500/ Fax: (022) 23665555/ info@radhagopinath.com

Mumbai, Maharashtra — Shristhi Complex, Mira Rd. (E), opposite Royal College, Dist. Thane, 401 107/ Tel. (022) 28454667 or 28454672/ Fax: (022) 28454981/ jagjivan.gkg@pamho.net

• **New Delhi, UP** — Hare Krishna Hill, Sant Nagar Main Road, East of Kailash, 110 065/ Tel. (011) 2623-5133, 4, 5, 6, 7/ Fax: (011) 2621-5421/ delhi@pamho.net; (Guesthouse) neel.sunder@pamho.net

• **New Delhi, UP** — 41/77, Punjabi Bagh (West), 110 026/ Tel. (011) 25222851 or 25227478 Noida, UP — A-5, Sector 33, opp. NTPC office, Noida 201 301/ Tel. (0120) 2506211/ vraja.bhakti.vilas.lok@pamho.net

Pune, Maharashtra — 4 Tarapoor Rd., Camp, 411 001/ Tel. (020) 26332328 or 26361855/ iyfpune@vsnl.com

Puri, Orissa — Bhakti Kuti, Swargadwar, 752 001/ Tel. (06752) 231440

Sri Rangam, TN — 103 Amma Mandapam Rd., Sri Rangam, Trichy 620 006/ Tel. (0431) 2433945/ iskcon_srirangam@yahoo.com.in

Surat, Gujarat — Rander Rd., Jahangirpura, 395 005/ Tel. (0261) 765891, 765516, or 773386/ surat@pamho.net

• **Thiruvananthapuram (Trivandrum), Kerala** — Hospital Rd., Thycaud, 695 014/ Tel. (0471) 2328197/ jsdasa@yahoo.co.in

• **Tirupati, AP** — K.T. Rd., Vinayaka Nagar, 517 507/ Tel. (0877) 2230114 or 2230009/ revati.raman.jps@pamho.net (guesthouse: iskcon_ashram@yahoo.co.in)

Udhampur, J&K — Srila Prabhupada Ashram, Srila Prabhupada Marg, Srila Prabhupada Nagar 182 101/ Tel. (01992) 270298/ info@iskconudhampur.com

Ujjain, MP — Hare Krishna Land, Bharatpuri, 456 010/ Tel. (0734) 2535000 or 3205000/ Fax: (0734) 2536000/ iskcon.ujjain@pamhho.net

Varanasi, UP — ISKCON, B 27/80 Durgakund Rd., Near Durgakund Police Station, Varanasi 221 010/ Tel. (0542) 246422 or 222617

• **Vrindavan, UP** — Krishna-Balaram Mandir, Bhaktivedanta Swami Marg, Raman Reti, Mathura Dist., 281 124/ Tel. & Fax: (0565) 2540728/ iskcon.vrindavan@pamho.net; (Guesthouse:) Tel. (0565) 2540022; ramamani@sancharnet.in

ADDITIONAL RESTAURANT

Kolkata, WB — Govinda's, ISKCON House, 22 Gurusaday Rd., 700 019/ Tel. (033) 24756922, 24749009

EUROPE (partial list)*

Amsterdam — Van Hilligaertstraat 17, 1072 JX/ Tel. +31 (020) 675-1404 or -1694/ Fax: +31 (020) 675-1405/ amsterdam@pamho.net

Barcelona — Plaza Reial 12, Entlo 2, 08002/ Tel. +34 93 302-5194/ templobcn@hotmail.com

Bergamo, Italy — Villaggio Hare Krishna (da Medolago strada per Terno d'Isola), 24040 Chignolo d'Isola (BG)/ Tel.

+39 (035) 4940706

Budapest — Lehel Street 15–17, 1039 Budapest/ Tel. +36 (01) 391-0435/ Fax: (01) 397-5219/ nai@pamho.net

Copenhagen — Skjulhoj Alle 44, 2720 Vanlose, Copenhagen/ Tel. +45 4828 6446/ Fax: +45 4828 7331/ iskcon.denmark@pamho.net

Grödinge, Sweden — Radha-Krishna Temple, Korsnäs Gård, 14792 Grödinge, Tel.+46 (08) 53029800/ Fax: +46 (08) 53025062 / bmd@pamho.net

Helsinki — Ruoholahdenkatu 24 D (III krs) 00180/ Tel. +358 (9) 694-9879 or -9837

• **Lisbon** — Rua Dona Estefânia, 91 R/C 1000 Lisboa/ Tel. & fax: +351(01) 314-0314 or 352-0038

Madrid — Espíritu Santo 19, 28004 Madrid/ Tel. +34 91 521-3096

Paris — 35 Rue Docteur Jean Vaquier, 93160 Noisy le Grand/ Tel. & fax: +33 (01) 4303-0951/ param.gati.swami@pamho.net

Prague — Jilova 290, Prague 5 - Zlicin 155 21/ Tel. +42 (02) 5795-0391/ info@harekrsna.cz

• **Radhadesh, Belgium** — Chateau de Petite Somme, 6940 Septon-Durbuy/ Tel. +32 (086) 322926 (restaurant: 321421)/ Fax: +32 (086) 322929/ radhadesh@pamho.net

• **Rome** — Govinda Centro Hare Krsna, via di Santa Maria del Pianto 16, 00186/ Tel. +39 (06) 68891540/ govinda.roma@harekrsna.it

• **Stockholm** — Fridhemsgatan 22, 11240/ Tel. +46 (08) 654-9002/ Fax: +46 (08) 650-881; Restaurant: Tel. & fax: +46 (08) 654-9004/ lokanatha@hotmail.com

Warsaw — Mysiadlo k. Warszawy, 05-500 Piaseczno, ul. Zakret 11/
Tel. +48 (022) 750-7797 or -8247/ Fax: +48 (022) 750-8249/
kryszna@post.pl

Zürich — Bergstrasse 54, 8030/ Tel. +41 (01) 262-3388/ Fax: +41 (01) 262-3114/ kgs@pamho.net

RURAL COMMUNITIES

France (La Nouvelle Mayapura) — Domaine d'Oublaisse, 36360, Lucay le Mâle/ Tel. +33 (02) 5440-2395/ Fax: +33 (02) 5440-2823/ oublaise@free.fr

Germany (Simhachalam) — Zielberg 20, 94118 Jandelsbrunn/ Tel. +49 (08583) 316/ info@simhachalam.de

Hungary (New Vraja-dhama) — Krisna-völgy, 8699 Somogyvamos, Fö u, 38/ Tel. & fax: +36 (085) 540-002 or 340-185/ info@krisnavolgy.hu

Italy (Villa Vrindavan) — Via Scopeti 108, 50026 San Casciano in Val di Pesa (FL)/ Tel. +39 (055) 820054/ Fax: +39 (055) 828470/
isvaripriya@libero.it

Spain (New Vraja Mandala) — (Santa Clara) Brihuega, Guadalajara/ Tel. +34 949 280436

ADDITIONAL RESTAURANTS

Barcelona — Restaurante Govinda, Plaza de la Villa de Madrid 4–5, 08002/ Tel. +34 (93) 318-7729

Copenhagen — Govinda's, Nørre Farimagsgade 82, DK-1364 Kbh K/ Tel. +45 3333 7444

Milan — Govinda's, Via Valpetrosa 5, 20123/ Tel. +39 (02) 862417

Oslo — Krishna's Cuisine, Kirkeveien 59B, 0364/ Tel. +47 (02) 260-6250

Zürich — Govinda Veda-Kultur, Preyergrasse 16, 8001/ Tel. & fax: +41 (01) 251-8859/ info@govinda-shop.ch

COMMONWEALTH OF INDEPENDENT STATES (partial list)*

Kiev, Ukraine — 16, Zorany per., 04078/ Tel. +380 (044) 433-8312, or 434-7028 or -5533

Moscow, Russia — 8/3, Khoroshevskoye sh. (mail: P.O. Box 69), 125284/ Tel. +7 (095) 255-6711/ Tel. & fax: +7 (095) 945-3317

LATIN AMERICA (partial list)*

Buenos Aires, Argentina — Centro Bhaktivedanta, Andonaegui 2054, Villa Urquiza, CP 1431/ Tel. +54 (01) 523-4232/ Fax: +54 (01) 523-8085/ iskcon-ba@gopalnet.com

Caracas, Venezuela — Av. Los Proceres (con Calle Marquez del Toro), San Bernardino/ Tel. +58 (212) 550-1818

Guayaquil, Ecuador — 6 de Marzo 226 and V. M. Rendon/ Tel. +593 (04) 308412 or 309420/ Fax: +564 302108/ gurumani@gu.pro.ec

• **Lima, Peru** — Schell 634 Miraflores/ Tel. +51 (014) 444-2871 **Mexico City, Mexico** — Tiburcio Montiel 45, Colonia San Miguel, Chapultepec D.F., 11850/ Tel. +52 (55) 5273-1953/ Fax: +52 (55) 52725944

Rio de Janeiro, Brazil — Rua Vilhena de Morais, 309, Barra da Tijuca, 22793-140/ Tel. +55 (021) 2491-1887/ sergio.carvalho@pobox.com

San Salvador, El Salvador — Calle Chiltiupan #39, Ciudad Merliot, Nueva San Salvador (mail: A.P. 1506)/ Tel. +503 2278-7613/ Fax: +503 2229-1472/ tulasikrishnadas@yahoo.com

São Paulo, Brazil — Rua do Paraiso, 694, 04103-000/ Tel. +55 (011) 326-0975/ communicacaomandir@grupos.com.br

West Coast Demerara, Guyana — Sri Gaura Nitai Ashirvad Mandir, Lot "B," Nauville Flanders (Crane Old Road), West Coast Demerara/ Tel. +592 254 0494/ iskcon.

guyana@yahoo.com

AFRICA (partial list)*

Accra, Ghana — Samsam Rd., Off Accra-Nsawam Hwy., Medie, Accra North (mail: P.O. Box 11686)/ Tel. & fax +233 (021) 229988/ srivas_bts@yahoo.co.in

Cape Town, South Africa — 17 St. Andrews Rd., Rondebosch 7700/ Tel. +27 (021) 6861179/ Fax: +27 (021) 686-8233/ cape.town@pamho.net

◆ **Durban, South Africa** — 50 Bhaktivedanta Swami Circle, Unit 5 (mail: P.O. Box 56003), Chatsworth, 4030/ Tel. +27 (031) 403-3328/ Fax: +27 (031) 403-4429/ iskcon.durban@pamho.net

Johannesburg, South Africa — 7971 Capricorn Ave. (entrance on Nirvana Drive East), Ext. 9, Lenasia (mail: P.O. Box 926, Lenasia 1820)/ Tel. +27 (011) 854-1975 or 7969/ iskconjh@iafrica.com

Lagos, Nigeria — 12, Gani Williams Close, off Osolo Way, Ajao Estate, International Airport Rd. (mail: P.O. Box 8793, Marina)/ Tel. +234 (01) 7744926 or 7928906/ bdds.bts@pamho.net

Mombasa, Kenya — Hare Krishna House, Sauti Ya Kenya and Kisumu Rds. (mail: P.O. Box 82224, Mombasa)/ Tel. +254 (011) 312248

Nairobi, Kenya — Muhuroni Close, off West Nagara Rd. (mail: P.O. Box 28946)/ Tel. +254 (203) 744365/ Fax: +254 (203) 740957/ iskcon_nairobi@yahoo.com

◆ **Phoenix, Mauritius** — Hare Krishna Land, Pont Fer (mail: P.O. Box 108, Quartre Bornes)/ Tel. +230 696-5804/ Fax: +230 696-8576/ iskcon.hkl@intnet.mu

Port Harcourt, Nigeria — Umuebule 11, 2nd tarred road, Etche (mail: P.O. Box 4429, Trans Amadi)/ Tel. +234 08033215096/ canakyaus@yahoo.com

Pretoria, South Africa — 1189 Church St., Hatfield, 0083 (mail: P.O. Box 14077, Hatfield, 0028)/ Tel. & fax: +27 (12) 342-6216/ iskconpt@global.co.za

RURAL COMMUNITY

Mauritius (ISKCON Vedic Farm) — Hare Krishna Rd., Vrindaban/ Tel. +230 418-3185 or 418-3955/ Fax: +230 418-6470

BHAGAVAD-GĪTĀ AS IT IS

The world's most popular edition of a timeless classic.

Throughout the ages, the world's greatest minds have turned to the *Bhagavad-gītā* for answers to life's perennial questions. Renowned as the jewel of India's spiritual wisdom, the *Gītā* summarizes the profound Vedic knowledge concerning man's

essential nature, his environment, and ultimately his relationship with God. With more than fifty million copies sold in twenty languages, *Bhagavad-gītā As It Is*, by His Divine Grace A.C. Bhaktivedanta Swami Prabhupāda, is the most widely read edition of the *Gītā* in the world. It includes the original Sanskrit text, phonetic transliterations, word-for-word meanings, translation, elaborate commentary, and many full-color illustrations. (Pocket version: translation and commentary only.)

Pocket	Vinyl	Hard	Deluxe
$4.50	$8.50	$9.95	$19.95
BGS	BGV	BGH	BGD

ORDER TOLL FREE 1-800-927-4152

For a free catalog call:

1-800-927-4152